THE SCIENCE OF WALKING

THE SCIENCE OF WALKING

Investigations into Locomotion in the Long Nineteenth Century

ANDREAS MAYER

Translated by Robin Blanton and Tilman Skowroneck
Revised and extended by the author

THE UNIVERSITY OF CHICAGO PRESS
CHICAGO AND LONDON

The University of Chicago Press, Chicago 60637
The University of Chicago Press, Ltd., London
© 2020 by The University of Chicago
Published 2020
Printed in the United States of America

29 28 27 26 25 24 23 22 21 20   1 2 3 4 5

ISBN-13: 978-0-226-32835-5 (cloth)
ISBN-13: 978-0-226-35248-0 (e-book)
DOI: https://doi.org/10.7208/chicago/9780226352480.001.0001

Originally published as *Wissenschaft vom Gehen: Die Erforschung der Bewegung im 19. Jahrhundert*, © 2013 S. Fischer Verlag GmbH, Frankfurt am Main.

Translated by Robin Blanton and Tilman Skowroneck. Revised and extended by the author.

The translation of this work was funded by Geisteswissenschaften International– Translation Funding for Humanities and Social Sciences from Germany, a joint initiative of the Fritz Thyssen Foundation, the German Federal Foreign Office, the collecting society VG WORT, and the German Publishers & Booksellers Association.

Library of Congress Cataloging-in-Publication Data

Names: Mayer, Andreas, 1970– author. | Blanton, Robin, translator. | Skowroneck, Tilman, translator.
Title: The science of walking : investigations into locomotion in the long nineteenth century / Andreas Mayer ; translated by Robin Blanton and Tilman Skowroneck ; revised and extended by the author.
Other titles: Wissenschaft vom Gehen. English
Description: Chicago : University of Chicago Press, 2020. | Includes bibliographical references and index.
Identifiers: LCCN 2019045111 | ISBN 9780226328355 (cloth) | ISBN 9780226352480 (ebook)
Subjects: LCSH: Walking—History—19th century. | Physiology—History— 19th century. | Science—History—19th century.
Classification: LCC QP310.W3 M3813 2020 | DDC 612/.044—dc23
LC record available at https://lccn.loc.gov/2019045111

♾ This paper meets the requirements of ANSI/NISO Z39.48-1992 (Permanence of Paper).

To the memory of my grandfather,
Oswald Alturban (1916–1998),
who taught me how to walk

Is it not really quite extraordinary to contemplate the fact that, since the moment man took his first steps, no one has asked himself why he walks, how he walks, if he walks, if he could walk better, what he does while walking, and if there is a way to impose rules on his walking, to change it, or to analyze it: questions that are pertinent to all the philosophical, psychological, and political systems that have preoccupied the world?
—Honoré de Balzac, *Théorie de la démarche* (1833)

Tullio was talking about his ailment again, which, indeed, was his principal pastime. He had studied the anatomy of the leg and the foot. With a laugh, he told me that in a brisk walk, each step takes no more than half a second, and that during that half second no fewer than fifty-four muscles are set in motion. My mind boggled and my thoughts immediately traveled to my legs, looking for this monstrous machine. I believe I found it. Naturally I did not find fifty-four separate devices, but rather a large and complex mechanism which lost all its orderliness as soon as I directed my attention to it.

Limping, I left the café, and my limp continued for several days. Walking had become a toilsome, even slightly painful endeavor. It seemed that this tangle of mechanisms had lost their lubricating oil, and now impeded one another in their motion. A few days later, an even worse evil befell me, about which I will also write, beside which the first paled. But even today, if anyone watches me as I move about, the fifty-four movements make fools of themselves, and I find myself in danger of falling down.
—Italo Svevo, *La coscienza di Zeno* (1923)

CONTENTS

A Recalcitrant Object

This book is about a science that does not exist, at least not in the conventional sense. *The Science of Walking* recounts the story of the growing interest and investment of Western scholars, physicians, and writers in the scientific study of an activity that seems utterly trivial in its everyday performance and yet essential to our human nature: walking on two legs. Since the late eighteenth century, theories of human walking have proliferated in various scientific, medical, and political contexts, often with the aim of controlling or improving the gait of certain populations. In an era that saw the rise of new forms of transportation and travel (such as the railway), the mode of transport regarded as the most "natural" of all acquired an entirely new status. Since the emergence of the industrial world, the "arts of walking" have been celebrated and cultivated by poets, philosophers, and intellectuals from Rousseau to Wordsworth and Thoreau, among the most familiar faces. In contrast, the history of the scientific theories and experiments that aimed to define the natural human gait with the aid of analytical techniques and measuring devices has attracted much less attention and remains largely to be written.

One reason for this relative neglect may be the fact that scientific attempts to measure gait in specific populations have been mostly identified with the project to increase efficiency and control over human subjects in Western societies, societies that have come to be characterized as driven by "mechanization" and the "acceleration of life."[1] "The indeterminacy of a ramble, on which much may be discovered," laments the author of a popular history of walking at the beginning of the twenty-first century, "is being replaced by the determinate shortest distance to be traversed with all possible speed, as well as by the electronic transmissions which make travel less necessary."[2] In a similar vein, a recent best-selling French book on the philosophy of walking praises strolling as an act of archaic and mystical rebellion in Western industrialized societies whereby subjects can liberate themselves in an even more radical way: "By walking, you can escape from the very idea of identity, the temptation to be someone, to have a name and a history. . . . The freedom in walking lies in not being anyone; for the walking body has no history, it is just an eddy in the stream of immemorial life."[3]

Such appraisals of the solitary walk speak all too clearly of the writer's or the philosopher's ardent wish to leave civilization, society, and history entirely behind. The poetics of walking, however, not only bears the traces of a long intellectual history ranging from Rousseau to Thoreau and Nietzsche that came to equate walking with free and radical thought. It also emerged in relation to two other registers: a *mechanical* conception of the anatomy and physiology of the body in motion, and a *semiotic* quest to identify the seemingly isolated walking subject with a specific character or type. Whereas the former has its roots in the iatromechanical medicine of the seventeenth century, which analyzed bodily motion using the laws of classical mechanics, the latter goes back at least to antiquity, when certain forms of walking provided essential clues to personality, social status, or even divinity. Its *locus classicus* is certainly the passage from the first book of the *Aeneid* in which Venus, disguised as a shepherdess, is at last recognized by her son thanks to her gait: "Et vera incessu patuit dea."[4] Although this physiognomic interest in different forms of walking and body postures can be detected in many cultures and epochs,[5] the semiotics of walking first acquired an important epistemic status during the eighteenth century with the emergence of the new *Science de l'homme* (science of man), which aimed to complement or oppose the prevailing mechanistic approach.

One of the main objectives of this book is to show not only that the walking body has a history but also that this history needs to be comprehended through a detailed reconstruction of the forms of knowledge that tried to capture bodies in motion. In other words, the approach pursued here aims to reframe the history of walking in terms of a history of scientific theories and practices of observation and experimentation within a wide range of disciplines (anthropology, physiology, orthopedic surgery, neurology, psychiatry) throughout the long nineteenth century. These various attempts to produce knowledge about walkers and their gaits were linked to moral ideals and political projects and thus need to be understood in their respective historical contexts.

That being said, it nevertheless remains difficult to integrate the history of the sciences of walking into the narratives that dominate the literature devoted to the history and sociology of the body. From stating that ever since its codification in conduct books during the sixteenth and seventeenth centuries, the ideal of the natural, upright gait was strongly shaped by moral and political values, it does not necessarily follow that its subsequent cultivation in new spaces such as the "promenades" was only an expression of forms of distinction between an emerging bourgeois walking culture and the nobility. Reducing the gait to a game of social distinction carries the risk of overlooking the specific forms of blending the social, the technical, and the biological that characterized physiologies of walking in the first half of the nineteenth century.[6] Similarly, to see in these new forms of scientific knowledge—following the theses of Michel Foucault's microphysics of power—merely the manifestation of a new and subtle form of disciplinary power would be to present knowledge making about human locomotion as just another instance of the normalization of bodies.[7] If scientific knowledge about body movement is restricted to performing exclusively normative functions, the epistemic and material specificities (and, one is inclined to add, the oddities) of this field of research are largely bypassed or erased.

Not all scientific investigations of human movement dovetailed with reform projects aiming to make bodies walk or march according to certain norms. Other approaches, notably those that would come to define the discipline of neurology, turned their attention exclusively to cases where walking simply failed or took on "irregular" or "bizarre" forms. The neurologists set themselves not only the task of building a new diagnostic knowledge of walking disorders but also the more challenging task of conceptualizing locomotion as an act of motor and sensory coordination. On an epistemological level, all researchers had to grapple with the problem of the *idiosyncrasy* of the gait, widely recognized as a marker of individuality. That anybody can be known by his or her gait is a topos that became a triviality in the nineteenth century, at least in the urban context of large cities, the terrain on which the scientific observers and theorists of walking usually operated. Possibly the best indicator of this development is in the famous opening of Balzac's novel *Ferragus* (1833), where the protagonist is able to identify the woman he adores by her gait alone: "By the way a Parisian woman wraps a shawl around her, and the way she lifts her feet in the street, a man of intelligence in such studies can divine the secret of her mysterious errand. There is something, I know not what, of quivering buoyancy in the person, in the gait [*Il y a je ne sais quoi de frémissant, de léger dans la personne et dans la démarche*]; the woman seems to weigh less; she steps, or rather, she glides like a star, and floats onward led by a thought which exhales from the folds and motion of her dress."[8]

Balzac's axiom does more than offer a modern and slightly trivialized version of Virgil's verse about Aeneas recognizing his mother Venus as a "true Goddess" by her step. It fulfils the dual function of attesting to the abilities of the astute male observer while also defining the qualities of the fleeting object of his observation. The walking woman is presented here as an enigmatic but utterly decipherable aggregate of parts. Balzac's ambitious *Théorie de la démarche* (1833) echoed the sense of many of his contemporary physiologists that the gait provided the greatest challenge to the "science of man" and that it could be solved with the help of scientific knowledge.

In many ways, Balzac's attempts to confront the scientific observer with the elusiveness of walking have informed the method that I gradually adopted myself during the gestation of this book. Following the understanding of most scholars of the nineteenth century, I came to put walking, along with other human activities located at the border of the physiological and the psychological,[9] into the class of recalcitrant objects. These objects resist to some extent practices of observation and experiment, but recognizing this resistance is not the same as endowing them with mysterious powers (an unknown energy, force, or "fluid") or equating them with the "ineffable." The realm that is designated here is possibly best captured by terms such as the *unconscious* (not in a psychological but in a purely physiological sense of body processes that are not conscious), or, to cite one of the conceptual innovations of early twentieth century anthropology, by Marcel Mauss's notion of "techniques of the body."[10] Although I consider his coinage of *body techniques* at the end of this book, I have resisted the temptation to subordinate the historically unfolding problem of turning walking into an object of knowledge to this or any other later theoretical formulation. Rather, it is *the progressive unfolding of the problem*, as it came to the fore in the endeavors of historical actors to solve it, that gives this book its structure.[11]

THE STRUCTURE OF THE BOOK

The four chapters of this book each foreground in turn a central component or set of practices in the process of knowledge making that became problematic during a given period. Chapter 1 centers on the various forms of *practical knowledge* that were characteristic of late eighteenth-century modes of relating to walking, often in an idealized fashion, as the most

perfect mode of human existence. Chapter 2 focuses on the first three decades of the nineteenth century, emphasizing the predicament of *observation* as it became most acute in the ambitious program of the French
Science de l'homme. In this context, practices of observation were often opposed to forms of experiment, following a well-known opposition
within the disciplinary development of anatomy and physiology but one
that took on specific forms in the case of human locomotion. Chapter 3
traces in closer detail the contexts for *experimentation* on the "human
walking apparatus" in Germany in the 1830s and some of its major critics in the following two decades. Chapter 4 analyzes the crucial developments in the study of locomotion that took place in France in the second
half of the nineteenth century and shows how the problem of *representation* came to the fore in an unprecedented way.

Many educated men and women of the eighteenth century considered
walking the ideal and most natural form of locomotion. Chapter 1 in-
quires into the reasons for this valorization by tracing the thought of some
of the many, mostly forgotten authors who wrote in praise of foot travel.
It asks to what extent this literature was inextricably linked to a critique
of mechanical means of transport. Popular philosophers, philanthropists,
and military reformers shared an interest in the study of the "admirable
mechanism" of the "natural" ways of walking and performing body movement. They based their theoretical views on an anatomical and physiological understanding of walking that was mostly derived from G. A. Borelli's
classical work *De motu animalium* (*On the Movement of Animals*),[12] but
they combined this approach from classical mechanics with a semiotics
that called for a "physiognomy" of individual or national characters. The
spread of Johann Caspar Lavater's theories about physiognomy throughout Europe is a famous instance of this latter development, so one has to
ask to what extent the reading of facial characteristics was also applied to
forms of gait and gesture.

Chapter 2 turns to the rise and increasing problematization of the observation of body movement within the French intellectual movement
known as the *Science de l'homme*. Some of its most prominent exponents,
such as the Montpellier physician Paul-Joseph Barthez, set out to refute
physiological approaches toward movement derived from Borelli's mechanics by positing that an "incalculable" play of vital forces occurred in
the muscles. While the Montpellier school favored practices of observation (not only of real bodies in motion but also the reviewing of classical
texts, literature, and art), the physiology of locomotion in the next decade
shifted gradually toward experimental practices. Chapter 2 asks whether

new laboratory practices, which sought to establish facts through animal vivisection, were successful in opposing the larger cultural framework of the anthropological physiologies of locomotion. The growing tension between a semiotic approach to walking, forcefully embodied in the medical reception of Lavater's physiognomy of the human body, and a new mechanics grounded on animal experiment and patient observation, played out in several polemical discussions among the Parisian medical establishment.

At the end of chapter 2, I argue that Balzac's irreverent intervention in his *Théorie de la démarche* (Theory of walking) must be understood in this context. The essay has often been read as a parody of the current scientific attempts to provide the foundations for a complete "science of man," but one has to ask whether it does not actually present a serious, even tragic account of human efforts to comprehend the act of walking through observation. Balzac's conception of a "social pathology" of the gait then, seems to lead to the skeptical conclusion that there may be no such thing as the "natural gait." It therefore becomes incumbent on the observer to record the entire spectrum of culturally shaped manifestations and deformations of walking made up of innumerable details and with his own eyes: an endless project, doomed to failure. When read in its proper context, Balzac's essay can serve as a cultural index of the epistemological predicaments of the human sciences of this period.

While chapter 2 details the ways in which the first physiologists of walking defined and executed practices of observation in the clinics or on the streets, chapters 3 and 4 are devoted to the rise of the experimental physiology of walking as it occurred first in Germany in the 1830s and subsequently in France during the second half of the nineteenth century. The treatise *Mechanik der menschlichen Gehwerkzeuge* (1836; Mechanics of the human walking apparatus) by the brothers Eduard and Wilhelm Weber seems at first glance to contradict Balzac's skepticism. In their painstaking attempts to subject the human gait to precise measurement and experimental control, these two scientists claimed to make a radical break with all former research. One must ask, however, whether their famous definition of the human "walking apparatus" (*Gehwerkzeuge*) as a perfect machine, following only the laws of gravity, was not in fact still strongly indebted to the epistemic practices and ideals of the late Enlightenment. The dangerous presence on German roads of new vehicles propelled by steam engines was an important context for the Webers' seemingly fantastical project to build a machine that walked on six or more legs. Another relevant context that shaped their experimental project lay in their extensive use of new optical-illusion devices. What, then, was the Webers' lasting contribution? Introducing precision instruments most commonly found in the observatories of the time into the study of human locomotion? Or combining them with novel forms of visual representation that they used to demonstrate their theory?

In the second half of the nineteenth century, a number of epistemic projects emerged that contributed to making the scientific study of locomotion visible at a larger cultural level. They are the subject of chapter 4. Among the most famous of these projects was the work conducted by physiologist Étienne-Jules Marey and his collaborators at the Collège de France in Paris from the early 1870s onward. They introduced self-recording instruments into the physiology of locomotion, an innovation that in many ways was presented as a scientific revolution. Marey's "graphical physiology" promised not only to make the "invisible" visible but also to provide the ultimate mechanical solution for the imperfections of human observers. With these new self-registering instruments, the mechanics of moving bodies found its most ambitious formulation. Marey strove to homogenize representations of movement with the claim that his devices were able to record a variety of "locomotion systems" (ranging from living organisms to inorganic objects and machines), and he staked a hegemonic claim with regard to every other form of representation of moving bodies whether in school physical education, in military institutions, or in works of art.

Since Marey's arsenal of apparatus and his "passion for the trace" has fascinated many later commentators, it seems necessary to understand his work in more than just the context of previous locomotion research. To what extent was his new outdoor laboratory—the "physiological station" where humans and animals were studied in motion—shaped by contemporary military institutions and artistic endeavors such as the circus? Whether Marey's approach to locomotion really constituted a sea change must be discussed in light of the controversies surrounding the correct depiction of horses in motion in the visual arts, in which aesthetic, epistemic, and political issues of representation were inseparably mixed.

In this context, it is necessary to turn more closely to the concrete sites where Marey's graphic and photographic recording devices for the study of human and animal locomotion were used, disputed, and sometimes rejected. What must one conclude from the fact that for most of the military horsemen who experimented at the physiological station, tracking hoofprints remained their preferred method of studying animal gait? Similarly, we need to clarify the extent to which the practices of observation developed by physicians were indebted to a different model: a distinctive set of clinical traceologies where observational skills in listening to a patient's footsteps or deciphering their footprints mattered. Although the relationship of this model with what Carlo Ginzburg has identified as the "evidential paradigm" of clues and traces seems evident, it is necessary to understand their different epistemological implications, especially when the focus widens from the individual case to the ambition to provide foundations for a classification of clinical types.[13]

The book closes by briefly considering the repercussions of the physiology of locomotion for epistemic formations in the human sciences that emerged in the first decades of the twentieth century. Its traces are detectable both in Sigmund Freud's psychoanalysis and in Marcel Mauss's anthropological conception of body techniques. As they deploy different concepts and work in separate spheres, these two projects may not seem to have much in common at first. But it is significant that both insist on the epistemic primacy of subjective experience involved in self-observation and postulate the supremacy of the symbolical over the biological. In the twentieth century, studies of walking in the human sciences are marked by an increasing self-reflexivity, and they become largely separated from biomechanical approaches, which have drawn conclusions from the complexity of the act of locomotion. While biomechanics have developed sophisticated techniques for simulating bodies in motion, it appears that anthropological studies of walking are bound to deal both with the observation and the recording of the real and thus must confront the problem of human reflexivity.

CHAPTER ONE

Walkers, Wayfarers, Soldiers:
Sketching a Practical Science of Locomotion

In the epistolary novel *Julie, ou la nouvelle Héloïse* (1761), Jean-Jacques Rousseau contrasted the aristocratic, elegant world of carriage travelers with "another" social world whose inhabitants moved, either of necessity or by choice, on foot. "Those who travel on foot," Rousseau wrote, "do not belong to high society; they are Bourgeois, common men, people from another world, and one could say that a carriage is necessary not so much to be driven places as to exist."[1]

Rousseau was not the only observer of his age to see, in driving and walking, not merely two different modes of transport but two fundamentally opposed modes of existence. Cultural and literary historians have previously noted that in the second half of the eighteenth century, a broad reappraisal of walking occurred, giving rise to an Enlightenment "bourgeois walking culture" that strove to dissociate itself from the court culture of representation by defining walking as a practice through which to experience and describe both nature and society.[2] This new culture of walking distanced itself from aristocratic codes of movement and above all from the artificiality and languor of the public "promenades," where the nobility demonstratively elected to ride in luxuriously appointed coaches rather than actually walk.[3] A large number of Enlightenment philosophers, writers, and educators defined themselves in explicit opposition to the aristocracy's mode of existence.

As Rousseau's example makes clear, however, the social outlines of the eighteenth-century walker and the walking life were not sharply drawn. "Bourgeois, common men, people from another world": these hardly define a coherent social class. This difficulty also pervades the extensive travel literature of the period, whose titles promoted the *Promenade*, the

Spaziergang, and the *Fußreise.*[4] Such texts declared walking the best and most natural way to travel, they and encouraged a new sensibility to the aesthetics and poetics of the walking life. Travel books, however, did more than redefine walking as a specific practice in contrast to coach travel. They also began to outline human locomotion itself as the subject of a new field of knowledge, as yet heterogeneous, in which mechanical and semiotic approaches intertwined.

What follows is an examination of these early efforts to turn the natural human gait into an object of knowledge, beginning with a definition of the anthropological points of departure for the enterprise, as these appeared in contemporary books about foot travel. Although the body of knowledge that grew up around walking drew on scientific theories, it was also practically focused, firmly oriented toward concrete places and activities. It circulated on the first semipublic promenades, where walkers discovered a new field of observation; in the schools of the Enlightenment philanthropinists, which developed a new pedagogy of walking; and on the parade grounds, where soldiers practiced marching in step.

ROUSSEAU'S ANTHROPOLOGY OF THE SOLITARY WALKER

Although Rousseau never authored any actual travel books, he surely must count as the most prominent exponent of the ambulatory way of existence as an aesthetical and moral project. As a result, his writings (published posthumously in 1782) provided a program for those who chose to travel on foot. In a famous passage from the *Confessions,* he articulates a walker's self-understanding:

> The thing I regret most about the details of my life the memory of which I have lost is that I did not make journals of my travels. Never have I thought so much, existed so much, lived so much, been myself so much, if I dare to speak this way, as in these travels I have made alone and on foot. Walking has something that animates and enlivens my ideas: I almost cannot think when I stay in place; my body must be in motion to set my mind in motion. The sight of the countryside, the succession of pleasant views, the open air, the big appetite, the distance from everything that makes me feel my dependence, from everything that recalls my situation to me, all this disengages my soul, gives me a greater audacity in thinking, throws me in some manner into the immensity of beings in order to combine them, choose them, appropriate them at my whim without effort and without fear. I dispose of all

nature as its master; wandering from object to object my heart unites,
identifies with the ones that gratify it, surrounds itself with charming
images, makes itself drunk with delightful feelings.[5]

Here, the solitary walker emphatically turns his back on the city, that em-
bodiment of civilization and corruption. On foot, he could appreciate the
beauty of nature and unravel its laws, while, at the same time, achieving self-awareness. Moving one's body becomes a central condition for
humans to understand and absorb the world around them. The epistemic
act, then, is intrinsically linked to a specific attitude. Rousseau envisions
himself in a kind of philosophical primal scene: he is the radical, fear-
less thinker who will rediscover the peripatetic methods of classical philosophy for the modern world. His roving gaze spurns the instrumental
apparatuses and classificatory systems of the natural sciences. By setting
himself free to move, seemingly purposelessly at first, he navigates toward
a complete understanding of the world and the laws that govern it.

In Rousseau's account of the nature stroll hinted at here, a poetics of
sorts, which would yield multiple expressions in the age of the French
Revolution and the early Romantic era,[6] there also dwells a fantasy of supremacy rooted in an anthropology that defined the upright human gait as
natural. The central elements of this anthropology may be found in Rousseau's early writings, particularly the *Discours sur l'origine et les fonde-
mens de l'inégalité parmi les hommes* (1755) and the educational novel
Émile (1762).[7] In the *Discours*, Rousseau finds it essential to show that
man, from his earliest origins, was a biped: "I will suppose him to have
been formed from all time as I see him today: walking on two feet, using
his hands as we use ours, directing his gaze over all nature, and measuring
with his eyes the vast expanse of the heavens."[8] With this proclamation,
Rousseau downplays multiple testimonials to the existence of people who
walked on all fours, familiar to him from stories of the so-called Hotten-
tots and the Caribs of the Antilles, as well as accounts of "wild" human
children raised by animals, preferring to follow, at least in broad strokes,
Buffon's portrayal of man in his *Histoire naturelle* of 1749 as sovereign
over nature and all living things.[9]

This did not prevent Rousseau, however, from developing his critique
of civilization and the sciences by drawing on reports from ethnographic
travel literature of the astonishing power and dexterity of the "noble savage."[10] The empirical specifics he cites are less important than the far-
reaching theoretical argument he derives, which culminates in a critique
of the very tools and machines of civilization:

Since the savage man's body is the only instrument he knows, he em-
ploys it for a variety of purposes that, for lack of practice, ours are incapable of serving. And our industry deprives us of the force and agility that necessity obliges him to acquire. If he had an axe, would his
wrists break such strong branches? If he had a sling, would he throw a
stone with so much force? If he had a ladder, would he climb a tree so
nimbly? If he had a horse, would he run so fast? Give a civilized man
time to gather all his machines around him, and undoubtedly he will
easily overcome a savage man. But if you want to see an even more un-
equal fight, pit them against each other naked and disarmed, and you
will soon realize the advantage of constantly having all one's forces at
one's disposal, of always being ready for any event, and of always carrying one's entire self, as it were, with one.[11]

Rousseau's anthropology thus regards the body as humankind's primary,
original instrument. Human beings are supposed to understand their own
hands, feet, and senses as an arsenal of natural "machines." He contrasts
these machines to the other tools developed by humans to understand and
master the natural world (tools that include, e.g., horses raised as livestock
or used as pack animals). In Rousseau's brand of Cartesian dualism, an
animal is a *machine ingénieuse*, an "ingenious machine," that survives
through natural instinct. A human, in contrast, operates the machine of the
body in the awareness of freedom of action.[12] In *Émile*, Rousseau's familiar
maxim to "always carry one's entire self with one" becomes the starting
point for an extensive critique of experimental physics and its instrumental
apparatus. Rousseau suggests that using instruments to measure distance
and weight will inevitably weaken the sensory functions of the human organs: "The more ingenious our apparatus, the coarser and more unskillful
are our senses. We surround ourselves with tools and fail to use those with
which nature has provided every one of us."[13] His advice to Émile's teacher
is to encourage the fictional pupil to use his physical senses in science lessons—in a long, laborious *physique expérimentale*[14] that sought to confront
the world unprotected—without the aid of books or measuring devices.

Within such an educational program as this, walking trips were man-
datory. Only on foot could an aspiring scholar acquire both the aesthetic
sensibility to appreciate nature properly, and an intellectual understanding of its workings:

To travel on foot is to travel in the fashion of Thales, Plato, and Pythag-
oras. I find it hard to understand how a philosopher can bring himself

to travel in any other way; how he can tear himself from the study of the wealth which lies before his eyes and beneath his feet. Is there any one with an interest in agriculture, who does not want to know the special products of the district through which he is passing, and their method of cultivation? Is there any one with a taste for natural history, who can pass a piece of ground without examining it, a rock without breaking off a piece of it, hills without looking for plants, and stones without seeking for fossils?[15]

"DRIVING IS IMPOTENCE; WALKING IS POWER": IN PRAISE OF FOOT TRAVEL

Rousseau was not the only thinker responsible for the rising appeal of foot travel in the second half of the eighteenth century. The descriptions of the Swiss naturalist Horace-Bénédict de Saussure, who published the first volume of his *Voyages dans les Alpes* in 1779, inspired other budding natural philosophers to make long study tours on foot. A young Georges Cuvier cited both Rousseau and Saussure when he undertook an eight-day walking tour of Swabia with a group of German students in 1788.[16] Ambulatory exploration of a terrain was, to be sure, essential to the specialized observational practice of geologists and botanists, but in the early Enlightenment, foot travel was also cultivated more generally among a small group of educated people:

> The mineralogist and botanist must travel on foot; their studies require it. In Germany, however, walking tours have also become the fashion with other persons who, though they are not mineralogists or botanists, have nevertheless found walking very convenient and comfortable. These people may walk more slowly; but one enjoys the journey twice as much, learns better the natural beauties of the countryside, and works up a glorious appetite. The walker meets with many small and sometimes extremely pleasant incidents and adventures, whereas for the hasty travelers in their coaches everything flashes past as in a peep-box.[17]

It is no accident that this description, taken from one of the first commercial travel guides distributed in Germany, contrasts the benefits of the walking tour with travel by coach. Walkers typically portrayed themselves to their readers as solitary travelers, pointedly declining rides in "machines," as coaches and carriages were frequently dubbed.[18] By turning their backs on society and its machines, foot travelers painted stylized pictures of themselves as quintessential mavericks, challenging others to follow in their footsteps. The walking tour was neither a specific genre nor a coherent program of social critique.[19] More than anything, it was a gesture of self-segregation from the organized and regulated system of coach travel.

The postal coach system at this time encompassed both the *Ordinari-Post*, which offered regular service on a number of routes, and the faster and more expensive *Extra-Post*. In the mid-eighteenth century, the coach system expanded to include passenger traffic in many parts of Europe, making travel time a more or less calculable quantity.[20] Passenger itineraries were dictated by the internal elements of the system: its stations (the so-called *Relais* or *Posten*) and signals (such as the post horn). Coach travel thus obeyed the specific strictures of a new principle of time efficiency—although not every stage of the journey was equally accelerated. Thanks to improvements to the road networks, horse-drawn carriages toward the end of the century could travel at ever higher speeds, but although driving time may have gotten shorter, it was punctuated by very long breaks at the stations. The time saved by traveling faster was always offset by the time lost in changing horses, paying fees, or loading and unloading baggage.[21]

The long walking tour emerged as a way of practicing walking that turned its back on this new, supposedly more efficient form of travel. The journalist Johann Kaspar Riesbeck ridiculed texts by travel writers who rode with the *Extra-Post* "in their tightly closed carriages."[22] Such remarks targeted figures such as the Enlightenment writer Friedrich Nicolai of Berlin, whose *Beschreibung einer Reise durch Deutschland und die Schweiz im Jahre 1781*, a work that spans thousands of pages, begins with a detailed description of the author's traveling carriage, which he likened to a "comfortable dwelling." Nicolai also advises travelers to plan every extended journey in meticulous detail.[23] He himself used a *Wegmesser*, a special kind of odometer, mounted on the wheels of his carriage to calculate his route more exactly. The *Wegmesser* had been constructed especially for him by the mechanician Peter Friedrich Catel, and Nicolai praises its benefits at length in an extra appendix to his book (fig. 1.1).[24] Instruments for measuring distance traveled had existed since antiquity. Now, as new modes of transport became more popular, refined versions of these devices allowed travelers to measure the approximate distances between the cities they visited, and tables of these distances appear in the first published travel guides.[25] Thus, the *Wegmesser* made Nicolai's carriage into a measuring instrument; he saw the carriage as a tool to

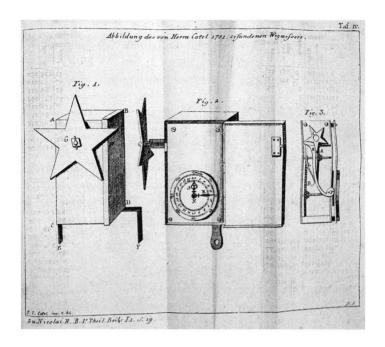

Figure 1.1 *Top*, Peter Friedrich Catel's odometer. *Bottom*, Zürner's odometer. From Friedrich Nicolai, *Beschreibung einer Reise durch Deutschland und die Schweiz im Jahre 1781.*

objectively define the actual routes he had traveled, and he praised its use-fulness in improving the "usually still quite inadequate maps."[26]

The polemics against the luxuriously appointed *Extra-Post* and its attendant route measuring can be traced, once again, to Rousseau, who called the carriage a "tightly closed cage" that interposed itself as a bar-rier between people and nature and promoted softness and decadence.[27] Many walkers modeled themselves on Rousseau's fictitious student Émile, who would "never enter a post-chaise."[28] Like "true knights-errant," they cultivated a slow, aimless manner of walking that deliberately rejected measurement and calculation.[29] Some of them undertook ambitious lit-erary projects that pushed readers to learn to see in a new way as well. The Hamburg publicist Jonas Ludwig von Heß, for example, responded to Nicolai's much-read travel reports with a seven-volume account of his own travels, *Durchflüge durch Deutschland, die Niederlande und Frank-reich* (1793–1800). The term *Durchflüge*, "fly-through," is definitely ironic. Heß is not concerned with speed but rather with the walker as a "nestless being," flexible and free. While acknowledging the inevitable "insufficien-cies" in his travel reporting, he still claims to be a more impartial and more precise observer than Nicolai, secure in the knowledge of the greater proximity of walkers to the objects they observe.[30] For Heß, the coach was no vehicle of the Enlightenment, as it appears, for example, in a 1793 cop-perplate engraving by the Berlin illustrator Daniel Chodowiecki in a land-scape literally flooded in sunlight (fig. 1.2).

A favorite complaint of walkers was that the coach made it harder to appreciate natural beauty owing to its erratic speed, the way it restricted the field of vision of its passengers, and not least of all, the company in-side. On a coach tour of England in 1782, Karl Philipp Moritz bemoaned the "always fragmented and interrupted view . . . which made me wish for a quick liberation from this rolling prison."[31] Moritz, a would-be *Ori-ginalgenie*, preferred to walk the English country roads with his Milton. Only reluctantly would he enter a coach, and then chiefly to avoid be-ing mistaken for a person of the lower classes, something he also never tires of decrying: "A pedestrian seems to be a legendary beast here, gawked at, pitied, held in suspicion, and avoided by everyone he encoun-ters."[32] In the "land of horse and carriage,"[33] anyone who would eschew a coach and "undertake a long trip on foot would be taken for a beggar or a scoundrel."[34] Moritz might be guilty of some exaggeration. Still, similar assessments certainly appear in other reports as well as in the earliest travel guides, including references to the numerous inconveniences that a

Aufklärung.

D. Chodowiecki ins & sc.

Figure 1.2 Daniel Chodowiecki, *Enlightenment*. Copper
engraving from the *Göttinger Taschenbuch* (1791).
Bildagentur für Kunst, Kultur und Geschichte.

"dilettante foot traveler" would encounter in Germany.[35] The jurist Heinrich Ludwig Christian Böttger even proposed a "uniform for foot travelers" to stem the "sad and inhuman prejudice" against walkers that was a "source of so much misery."[36]

In the end, foot travelers could not do entirely without driving; on certain routes it repeatedly proved a necessary evil. Walkers who wanted independence from the postal transport system and did not want to indulge in the luxury of the *Extra-Post* could hire a "Carriole," or cabriolet, a two-wheeled conveyance usually pulled by a single horse. Riding in cabriolets was a rough and jolting experience, often presenting a stark contrast to the natural beauty travelers saw around them. The school superintendent and educator Gerhard Anton Ulrich Vieth of Dessau, for instance, wrote to his parents in 1790,

> In Naumburg I took a cabriolet, a miserable thing, and an old stiff horse that had never been hitched to a wagon before; now imagine towering mountains and rocky ravines, and you will have some idea of our pilgrimage. To paint one scene for you, going uphill usually looked like this: I myself at the front, pulling the horse by its halter; the horse biting like the devil every time you took hold of it, and planting one on my arm, though causing no more harm than a mark on my sleeve. Behind us the cabriolet; and bringing up the rear, my companion, pushing with all his might. All this to the accompaniment of a storm of whiplashes and unceasing shouts. Apart from this—which admittedly made the rest all the more piquant—everything was indescribably beautiful.[37]

Tales of accidents in reluctantly entered cabriolets were stock-in-trade for authors who favored walking as they strove to paint sufficiently dire pictures of driving on impassable roads (fig. 1.3). After fleeing the Revolution in 1791, the French royal officer Jacques-Louis de Latocnaye (1767–1823) explored the British Isles in exile on foot for ten years; in 1802 he moved on to Sweden and Norway. He writes of a fall from a cabriolet that put a sudden stop not only to his aesthetic appreciation of the countryside but also to his continued mobility. "When I fell, my leg got stuck between two tree branches, slid down between them, and I landed on the ground with a sprained and dislocated knee, incapable of moving. Alone, in such a remote country, unable to walk, what would become of me? So full of horror seemed now the beautiful landscape and the beautiful Ångerman River!"[38]

Figure 1.3 Johann Heinrich Ramberg, *Unfälle zu Pferde und zu Wagen* (1806). Landesgalerie des Niedersächsischen Landesmuseums Hannover.

What is perhaps the quintessential carriage scene appears in a report by another military man and a famous walker: Johann Gottfried Seume (1763–1810), who embarked on his famous "stroll" to Syracuse carrying only a simple sealskin knapsack packed with clothes, shoes, and a small travel library. Precarious weather and the "dangerous swamps" between

Ferrara and Bologna forced Seume into a horse-drawn cabriolet, but that only made his voyage riskier and more difficult.

> We paid well and rode badly, and would have ridden even worse, had we not from time to time walked some of the worst stretches on foot. The first few hours from Ferrara went passably, but then the carriage sank in down to the axle. . . . We stepped out and struggled forward on foot, while for the empty carriage things got worse and worse. First *one* horse fell, and when it stood up again, the other one fell, and a few hundred steps later both fell and thrashed exhaustedly on the muddy clay ground."[39]

Finally, Seume says, he himself pulled the carriage out of the mud using all the strength of his "physical being." To the amazement of his fellow travelers, he then guided them across the impassable patch, giving them a "truly high opinion" of his "strength and agility."[40]

The dangers of coach travel were well known to every traveler of the time. Seume makes of them a dramatic story whose moral is that the strength and dexterity of the walker surpassed the performance of the horse-drawn carriage. For many nineteenth-century critics and readers, Seume's *Spaziergang* was an "audacious and ingenious undertaking."[41] It voiced the concerns of a generation of educators and popular philosophers who saw the upright human gait not only as evidence of human supremacy over animals but also as a condition of humanity itself. In his critique of progress in technologies of transportation, Seume thus had humanitarian aims. Animals and machines were expedients of which humans ought to divest themselves. Only by walking unsupported, on their own two feet, could they win back their autonomy and freedom. Playing on the multiple uses of the German verb *gehen* and its derivatives, Seume writes,

> I consider walking [*den Gang*] the most honorable and most independent thing in man, and I believe that everything would go [*gehen*] better if people would walk more. Almost everywhere, people can hardly rise to their feet and stay on them, simply because we drive too much. For anyone who sits too much in carriages things cannot go [*gehen*] well. We feel the truth of this; it seems unavoidable. When a machine stops moving we still say, as if we were still very much actively involved: "It does not run [*gehen*]." . . . The instant we sit down in a carriage, we distance ourselves by a few degrees from our original humanity. We can no longer look anyone in the eye as we ought, steadfast and

true: we are compelled to do too much or too little. Driving is impotence; walking [gehen] is power.[42]

AN EDUCATION IN WALKING: THE GYMNASTICS
OF THE PHILANTHROPINISTS

The ideal of the upright gait as the most natural form of locomotion also dominated the pedagogical and medical literature, where writers increasingly spoke out against the machines and contraptions that the upper classes had used since the late sixteenth century to teach their children to walk. The proverbial example is the so-called leading strings that were sewn onto the clothing of young children (fig. 1.4).[43] These, along with the go-carts and children's stays that were used to prop children up during their earliest years, now became the targets of both medical and moral criticism. It was Rousseau who delivered the final verdict: "There is nothing so absurd and hesitating as the gait of those who have been kept too long in leading strings when they were small. This is one of those observations which are considered trivial because they are true, and indeed true in more than one sense."[44] (Here Rousseau refers to the fact that "to be in leading strings" was a widely used idiom.)

This moral criticism of bondage (both physical and intellectual) was closely tied to the increasingly prevalent view in medicine that the "method of teaching children to walk by means of leading strings and go-carts" was "highly objectionable."[45] In 1778, the medical doctor and scholar Johann Georg Krünitz described the method in drastic terms in his Ökonomisch-technologische Encyclopädie:

They [children on leading strings] almost always appear red in the face—or brown, to be sure, or blue. Their heads and their whole bodies are bent forward. The leading strings are securely attached to the back and sides of the bodice, engirdling the shoulders completely. They have to carry the entire weight of the child; the child himself can hardly touch the ground. Because the child constantly leans on the strings with his chest, the chest—which, of all parts of the body, ought to be completely free to move without constraint!—is violently compressed by the front of the bodice, preventing the proper circulation of blood through that area; meanwhile, in the head, the circulation is less restricted, so all the blood accumulates there and dilates the blood vessels. Yet many of the ill effects that typically arise from such an accumulation of blood in the vessels of the head are quite well-

Figure 1.4 Peter Paul Rubens, *Rubens, His Wife Helena Fourment (1614–1673),*
and their Son Frans (1633–1678). Helena holds her son by a leading string.
From Dagmar Feghelm and Markus Kersting, *Rubens: Bilder der Liebe.*

known. The bent posture of the child that results from the backward
pull on the bodice forces the shoulder blades to overlap toward the
spine, lifts the shoulders, and can even dislocate the vertebrae. This
last is particularly to be feared if, as very often happens, the child is
only held by one string, and allowed to walk around it in a circle. As

the child's body is not planted firmly on the feet, the feet develop their full strength only very slowly, and they can even be twisted or sprained by being dragged across the ground.[46]

In newly founded German educational institutions in Dessau and Schnepfenthal, the educational reformers known as philanthropinists became the first to make *Leibesübungen*, "physical exercises," a central part of their curriculum both in theory and in practice. They, too, systematically rejected such devices as leading strings and go-carts. The German pastor and educator Christian Gotthilf Salzmann, who founded the philanthropinist school at Schnepfenthal in Gotha in 1784, advised mothers against using go-carts and recommended allowing babies to crawl before they learned to walk.[47] Salzmann's colleague Johann Christoph Friedrich Gutsmuths authored the first systematic manual of physical education, *Gymnastik für die Jugend* (1793; Gymnastics for youth), in which he rejects all the "artificial aids" for young walkers that were customary in the nursery: "They impart an entirely incorrect posture to the body and entirely incorrect habits to the young student."[48] Following Rousseau's maxim, Gutsmuth held that children learning an upright gait should mainly be left to rely on their own resources and their innate sense of their body's strength: "Nothing more is required to master this skill than what Nature has provided."[49]

Teaching children to walk was a job not for machinery but for teachers. However, instilling proper gait still meant offering frequent correction. During "walking practice," pupils had to walk back and forth one at a time in front of their fathers or a teacher and then listen to a critique from their fellow students:

[The educator] should frequently make each [pupil] walk before the other onlookers, who will be their judges, in various directions; first away from the group, then toward it, and finally back and forth in front of it; once at an easy pace, once briskly, and once fast. These variations will cast sufficient light on any person's gait. Afterward, everyone shall give his opinion regarding any faults that he observed; and since even youngsters possess quite correct instincts in this area and can easily perceive even the slightest caricature, errors will not easily go undetected.[50]

The philanthropinists took some key features of their definition of natural gait from the old aristocratic conduct books, a genre dedicated since the publication of Castiglione's *Libro del Cortegiano* in 1528 to describing the perfect courtier,[51] especially the requirements for an upright posture and an easy, elegant habit of movement.[52] Given that the first philanthropinist institutions set out to educate a predominantly aristocratic elite, this is hardly surprising. Crucially, however, the philanthropinists expressed these attributes of the courtly habitus as a universally applicable formula, giving them a new and expanded significance: "A truly seemly gait in a man arises only from the expression of physical strength and dexterity, for these are the qualities that create an agile, light, elastic, yet firm and manly walk; automatically produce an upright posture and a light and natural rotation of the body; and prevent a sunken chest, sagging shoulders, a bent neck, and limply swaying arms."[53]

In their efforts to develop a new definition of correct gait, the philanthropinists found additional fodder for criticism in the old methods of the dancing masters.[54] Certainly the new textbooks in physical education devoted ample space to the arts of fencing, dancing, and equestrian vaulting—alongside the classic triad of running, jumping, and walking—in their efforts to revive the gymnastic tradition of ancient Greece for the late eighteenth century.[55] But the expertise of the dancing master was now strictly circumscribed. Thus, Gutsmuths, for instance, could concede that the art of dancing might possibly contribute to "a good, decent manner of walking" but still advise categorically against "desiring to carry over any kind of dance step or posture to the everyday carriage of the body when walking."[56] Vieth, in his *Versuch einer Encyclopädie der Leibesübungen* (1795), praises the transformation that a dancing master could effect in his students but only when his methods were consistent with the natural laws of human anatomy. The "cruel school" that forced the feet into the dancer's turnout using such apparatuses such as the *tourne-hanche*, or "hip-turner," was rejected in favor of simpler exercises (fig. 1.5).[57] In the same vein, Vieth also rejected "affected manners and stilted postures" not only on aesthetic grounds but also on medical ones.[58]

The reasoning of these popular philosophers was symptomatic of a late-Enlightenment shift in thinking that set old, artificial ideals of posture, gesture, and foot position against a new anatomical, physiological view of the body as an "admirable machine" with "the innate capacity to perform an infinite variety of movements."[59] Thus, Vieth dismisses dance texts as long-winded, confused works that "affect a mathematical methodology, invoking geometry wherever a few straight or curved lines appear. O Father Euclid! Forgive them, for they know not what they do!"[60] He

Dighton del.

Mr. Deputy Numskull-taking a Lesson at Mr. Clumsy's dancing School.

Figure 1.5 *Boîtes* and *tourne-hanche*. Caricature by Robert Dighton from *The Town and Country Magazine; Or, Universal Repository of Knowledge, Instruction, and Entertainment* (1785). Victoria and Albert Museum, London.

gives natural movement a mathematical definition of a completely new kind, one based on the mechanics of moving bodies:

> The simplest and most natural manner of moving the body from one place to another is by walking, which consists in the following: the supports that bear up the body are moved alternately one in front of the other, and the body's center of gravity is shifted first above one [support], then the other. Walking, therefore, is partly a consequence of the musculature and its action, and partly of gravity: because as the center of gravity is transferred from one foot to the other, it actually *falls* through a small space, only to be caught immediately by the succeeding foot.[61]

A "continually *prevented* fall":[62] this way of explaining upright gait found widespread currency in the late eighteenth century thanks largely to popularizations of the fundamental principles of Giovanni Alfonso Borelli's famous treatise *De motu animalium* (1680–1681) in works such as Johann Gottlob Krüger's widely disseminated *Naturlehre* (1748).[63] According to Krüger's appraisal, Borelli's writings contained some "excellent truths, but also some curious fantasies, and it has been the peculiar lot of his book on animal locomotion to be praised by all, seen by some, read by few, understood by even fewer, and judged with sense by very few indeed."[64]

In this mechanistic physiology, the body was more than merely a well-arranged, efficiently designed machine. In the execution of its movements, it was also believed to achieve an "excellent" aesthetic effect. Krüger's reasoning, which was echoed by Vieth, is typical of this view: the attachment position of the muscles close to the joints (the *Ruhepunkte* [fulcra] of the machine) is not a "squandering" of energy since it serves the greater purpose of shortening the movement of the muscle. This avoids a "shapeless" body form and also allows for the "fastest possible movements."[65] It was perhaps in the late eighteenth-century military, in the revision and modernization of traditional drill techniques, that this new economy and aesthetics of walking and movement, which drew purely on the fundamental laws of mechanics, found its truest expression.

TACTICAL SPACES: REGULATING THE SOLDIER'S STRIDE

"The most important part of a soldier's training," proclaimed the Prussian general Friedrich Christoph von Saldern in 1781, "is marching: he must be perfectly practiced at it, in companies both small and large; and when

drilling in large formations, he must be able to take care of himself without being prompted."[66] In eighteenth-century armies, the codification of
military marching became a site for the construction of new knowledge.
This is reflected especially clearly in contemporary texts on tactics and
maneuvers, which recommended modifying the regulation stride lengths
for the infantry that had been prescribed in older drill manuals. These efforts at quantifying the "never-ceasing, never-changing movement of the
body"[67]—manifestations of a new *Krieges-Wissenschaft* (war science) that
grew steadily in prestige in the German states throughout the latter third
of the eighteenth century[68]—drew chiefly on the same mechanics of gait
on which gymnastics theoreticians such as Vieth also relied.[69] Its principles were simple, consisting in the calculation of the body's center of
gravity and its base of support, that is, the space above which the body
had to balance when standing or walking, and "laws of nature, which men
observe all the more strictly, for the certainty that a breach is synonymous
with the inevitable punishment of a fall."[70]

Appealing to the naturalness of a gait underpinned by mechanical
laws, new exercise handbooks and treatises on tactics advocated for a revision of the older drill books, which had instructed soldiers in convoluted
movements for handling weapons and prescribed a marching posture
with rigid torso and knees held stiff.[71] The early eighteenth century had
sought to achieve a mechanical coordination of the infantry. Exemplified
in the Prussian drill regulations, it called to mind a well-oiled clockwork,
one that was showcased in public processions and parades. To that end,
it was paramount that each soldier "assume a proper air, that is, hold his
head, body, and feet quite easily, and pull in his stomach," as the 1743
drill manual for the Prussian infantry put it.[72] The shouldering and carrying of the firearm (spelled out in a series of complicated grips and moves);
the straight stance; marching forward and backward in line formation—
all these movements were chiefly of a representative character and came
under increasing criticism from military theorists from the middle of the
century onward.

In particular, the Seven Years' War (1756–1763), whose uneven, obstacle-
riddled battlefields made line infantry tactics more difficult, spurred a development, beginning in France, toward more dynamic and flexible maneuver tactics. The sequences of movements laid out in earlier rule books
were criticized as artificial and impractical. Marching rules that grew out
of mechanical principles of locomotion, in contrast, were seen as an expression of natural law and as such increasingly became the focus of theoreticians. Deriving marching pace from the simple mechanical foundation

of natural locomotion also appeared to solve the problem of how to achieve
a single, uniform pace in the face of individual, national, and class-specific
differences in gait. In one of the most important late-Enlightenment treatises on military theory, the *Essai général de tactique* (1770), the Comte de
Guibert (1743–1790) writes

> Every human class, every nation, has its own gait, just as it has its
> own physiognomy. If one observes the gait of a Basque or a German;
> a Dutchman or a Provencal; someone raised in the cities or someone
> who lives in the country; a workman or an artist; one is bound to rec-
> ognize these differences; even in the gait of two brothers, born in the
> same climate and educated in the same profession, one will observe
> that one will lower the tip of his foot, and the other walk on his heel;
> one will walk heavily and slowly, the other with lightness and speed;
> these are all inevitable outcomes of the differences in their constitu-
> tions and characters and the mechanical folding and individual move-
> ments their legs have learned since childhood. The mechanism of loco-
> motion works similarly in every human being only with respect to one
> point only. In every person, the body follows the motion of the leg; the
> weight of the body is supported by the leg that stands upon the ground,
> and at the moment this leg is placed on the ground, the opposite foot is
> lifted to make the second step. As such, the principles of my drill step
> are correct and in agreement with nature.[73]

Guibert distinguishes between an at ease travel or route step (*la
marche de route*), which did not have to be practiced methodically, and a
strictly regulated drill step used for maneuvers, which should have "a uniform and collective mechanism" with a prescribed ordinary stride length
of "eighteen to twenty inches."[74] Finding a march step of sixty paces per
minute too heavy and cumbersome, Guibert increases "ordinary time"
(*pas ordinaire*) to eighty paces per minute. This put double time (*pas dou-
blé*), the "true military march,"[75] at 160 paces per minute and triple time
(*pas triplé ou de course*) at a full 200–250 paces per minute. Even before
they could be drilled to march, however, soldiers also had to be taught
to stand in a way "not at odds with the mechanical functioning of the
body."[76] Guibert disapproved of methods that would achieve this by more
or less forcible means: aligning soldiers' bodies against a wall, for instance,
or "the bizarre dress that half-crushes them, compressing all the joints."[77]
He prescribes a naturally straight body and head position with shoulders
back, "chest open,"[78] knees fully extended, "the two heels in a straight

line, two inches apart, not touching, and the feet turned out a little,"[79] following the tenets of anatomical mechanics, which defined the stability of
the body based on its line of gravity.

Once a recruit was comfortable adopting proper posture "on his own
and without compulsion, not as an exercise,"[80] training in the ordinary
march step could begin. The step was taught in two stages. First, the soldier was to "move the left leg forward, quickly but smoothly, with the
thigh turned outward a little; the foot moving flat and parallel to the
ground at an elevation of two inches, stopping when the left heel comes
level with the tip of the right foot." The movement came from the hip,
"the knee not stiff, but with a slight, easy bend."[81] On the count of "two,"
the soldier should "advance the left foot, with the body remaining straight
and following the movement of the leg. When, in the second stage, the foot
has moved forward by twelve inches, it is placed on the ground. The body
is carried forward to find itself almost entirely supported by the left leg,
and the right foot will rest lightly on the toe, with the right heel raised
and ready for the second step."[82]

To drill entire companies to march in step, Guibert proposes marking
off the exercise area with parallel strings to which "small markers of black
or red fabric" were attached. "In this way one can get the soldiers used
to lengthening their step to the required distance, marching in unison,
and maintaining distance between the ranks, since the wings of each rank
are required to reach the ends of the strings in a number of steps equal
to the number of the aforementioned markers."[83] An officer checked the
various march speeds with a stopwatch. Later, practice was moved from
the parade ground to open terrain, where instead of strings and markers,
officers who were "especially sure of their paces" directed the marching
from a central position. These officers also had to possess a quality known
as the *coup d'œil*, literally "stroke of the eye," that is, the ability to swiftly
size up a situation, including any particular challenges of the terrain. This
visual ability was one of the most important distinguishing qualities of
an officer.[84]

Guibert's recommendations were influential not only in France but
also in Germany, where they were used to revise German drill regulations.[85] Guibert likens the soldier not to an automaton but to a "living
statue" (*statue animée*), "always poised for action."[86] Similarly, most German drill books also emphasized the importance of "natural posture."
General Saldern, for example, writes in his *Taktische Grundsätze* of 1781
that a soldier needed to "stand straight underneath his rifle, but easy, so as
not to give the appearance of a marionette."[87] The training of a soldier—

Saldern calls it "Dressur"—thus emerged, at least in its ideal form, as a knowledge-driven process wherein the brute remaking of recruits into mechanized pieces in a great machine was replaced by the rational, pragmatic application of natural laws and *their* mechanisms. A company on the march became an apparatus of moving bodies that was more efficient and better adapted to the vagaries of combat and within which each soldier—using knowledge internalized through constant practice—acquired great individual mastery of all the march steps. "The various cadences and lengths of the paces must be so familiar to him that he will fall out of step only in the most difficult terrain and only when his strength fails him. He learns to fall into line quickly, on his own without being summoned, and with regular practice it is possible to get a large company to march just as easily and in just as orderly a fashion as a small one, which is always the main reason for these sciences."[88]

The creation of a disciplined body, one whose vital functions could be consistently deployed through precise calculations to achieve specific goals, was an ideal held not only by late eighteenth-century military tacticians but by other Enlightened reformers as well. In practice, however, implementing the simple laws of anatomical mechanics that underpinned the new drill books was neither smooth nor systematic.[89] A particularly controversial problem was how best to mark time so as to accustom marching soldiers to the proper beat. The Württemberg mathematician Franz Georg von Miller, for example, criticized Guibert's recommendations on this point:

> To understand how useless this instruction is, consider: when the command to march is given, a period of two to four ticks of the timepiece could easily pass before a man could begin moving from a standing position; or—and this will be still more common—before all the soldiers in a large unit will hear it. Moreover, how can a man with watch in hand, who is supposed to observe when the minute begins and ends even as he marches, now slower, now faster—how can this man, given how slowly the minute hand moves, execute his assignment with sufficient precision and not signal the end of the march a few paces too early or too late?[90]

To achieve a marching cadence of the correct speed, Miller recommended a method that did not rely on reading a watch face and then signaling but instead used specially constructed pocket watches. Each soldier was supposed to adjust his own step to the beat of the timepiece solely by ear.[91] Music was an even better tool; Miller called it "the most efficacious means of properly imprinting the rhythm of the march on the soldier: of interweaving it, so to speak, with his soul."[92] To indicate the correct beat to both musicians and marching unit, a simple pendulum was used. The pace to be adopted could be read from its oscillations. Miller and a friend, Jacob Friedrich von Rösch, a mathematician and captain, also conducted a series of walking experiments on a measured track in order to revise regulation stride lengths based on more accurate measurements.[93]

The efforts by Guibert and Saldern to correct older tactics by describing military marching with greater mathematical precision mark the beginnings of experiment-based investigation into the mechanics of gait. At the time, however, their work had only limited influence. As Miller's emendations to Guibert's method show, he mistrusted the *coup d'œil* and sought a method of communicating the marching beat that was purely auditory. But his suggestion that the infeasibility of taking exact readings from a measuring device might present a problem failed to jibe with the actual demands of military maneuvers, which often called for agility at the expense of precision. It also stood at odds with the literature on military theory, which lionized the great generals and their brilliant visual judgment. Thus, a review in the *Allgemeine Literatur-Zeitung* criticized Miller for his "pedantic investigations," for "over-thinking things," and for presenting his results according to "his own opinion" rather than invoking the expertise of famous "men of experience and insight."[94] Other critics found fault with Miller's use of "ordinary pocket watches" instead of watches with a second hand and with the "distracting" nature of march music.[95]

Miller's recommendation of the pendulum, however, fell on fertile soil. The physicist and military theorist Heinrich Johannes Krebs, who republished Saldern's *Taktische Grundsätze* in 1796, emphatically recommended the pendulum for training recruits in the slow *Schulmarsch* (schooling march): "since, even when marching to a watch with a second hand, it is hard to be sure that one has maintained a consistent, even pace; not to mention that it is impossible to expect that everyone who wishes to teach himself or others to march will possess a watch that tells the seconds, but it is easy to devise a pendulum under any circumstances."[96] Even so, the adoption of the pendulum owed less to the advantages of heightened precision and more to a highly desirable effect of demonstration and imitation that clearly abetted the cause of military gait training. In contrast to a clock face, the pendulum's oscillation suggested itself as an artificial model of the leg swaying to the laws of mechanics.

PHYSIOGNOMIES OF GAIT: A MORAL
SEMIOTICS OF HUMAN LOCOMOTION

The simple, intuitive mechanics of walking that found their clearest expression in the late-Enlightenment literature on military tactics were
also accompanied by a moral semiotics of gait. In the Enlightened military, it was taken for granted that a disciplined posture and march step
derived from natural laws were also indicative of soldierly "virtue."[97] And
it was no coincidence that Swiss pastor and physiognomist Johann Caspar
Lavater, in his *Physiognomische Fragmente* (1775–1778), sets readers the
challenge of matching a series of "important postures from the soldier's
world" to the "true character" expressed by each: "From the idealistic
majesty of the proudest general, completely conscious, not so much of who
he is but rather of how he appears or wishes to appear, down to the soldier
who walks the meanest streets—what a great and significant diversity!
Generals and officers, in all manner of imperious, worthy and unworthy,
natural and unnatural postures. Dexterity; rigidity; somnolence; affectation; insolence!"[98] (fig. 1.6).

First and foremost, Lavater's physiognomy involved the speculative,
theologically inspired deduction of character from the profile of the face.[99]
In the fourth and final volume of the *Fragmente* (1778), however, he casts
his net more widely to include clothing, voice, gait, gesture, and posture
as well. In the *Fragmente*, Lavater calls stance and gait "undisputedly
the most characteristic" personal traits. "Every virtue, every vice, every
strength and weakness," he writes, "has its own distinctive gait. A man
like Chodowiecki could sketch a thousand postures—ten thousand—each
one indisputably unique."[100] The book reproduces an array of drawings by
Daniel Chodowiecki, the popular Berlin illustrator, as well as images by
illustrator and entomologist Johann Rudolph Schellenberg. With them,
Lavater proffers to his audience a wide variety of social types: the "sweet-
talker" (*Süßling*), for example, and the "show-off" (*Windbeutel*), both variants of the "fop" or "dandy" (*Gecke*). As these "postures from the soldier's
world" show, Lavater's physiognomy relies almost entirely on the depictions of movement captured by his two illustrators. He himself adds only
the briefest commentary, in the form of moralizing vignettes, when he
does not leave the pictures to speak entirely for themselves.

The sparseness of his rhetorical apparatus proceeds naturally from
the purpose of his physiognomy, which was designed to promote "Menschenkenntnis und Menschenliebe" (insight into human nature and
philanthropy)—or, to put it more prosaically, for immediate daily use in

Figure 1.6 Daniel Chodowiecki, *Bedeutende Stellungen aus der Soldatenwelt.*
From Johann Caspar Lavater, *Physiognomische Fragmente*, vol. 4 (1778).

social interactions.[101] The observation that bodily movements and postures exhibited near-infinite variety and always varied from one individual to another inevitably raised the issue of how or whether such variety
could be expressed in terms of general scientific principles. Lavater con-
fronts this dilemma by postulating the existence of a harmony among all
observable movements. In practice, this meant paying extra attention to
concordant lines of face and body in the persons he observed, the shapes

of which lines he always interpreted as unequivocal manifestations of specific moral traits.[102] In a manuscript written for his friends, Lavater offered
advice on how to use this kind of physiognomic observation to penetrate
the art of disguise practiced by the female sex:

> If the gait of a woman is deadly, absolutely deadly—not merely dis-
> agreeable—but impetuous, crooked, without dignity—contemptuous,
> sidling—then be not tempted by her beauty, nor her wit, nor her trust.—
> Her mouth will be like to her walk and her conduct like her mouth,
> hard and false.. She will thank you for nothing that you do for her, and
> for the least thing you fail to do she will take terrible revenge—compare
> her gait and the lines of her brow, her gait and the creases around her
> mouth; you will be amazed by the harmony between them.[103]

The persuasive power of Lavater's arguments depended heavily on
Chodowiecki's engravings, a fact that did not escape Lavater's more fervent critics. One such critic, the Göttingen philosopher and physicist
Georg Christoph Lichtenberg, having published a number of polemics de-
nouncing the "physiognomy frenzy," launched a counterproject for which
he, too, secured the services of the Berlin illustrator.[104] Like other critics,
Lichtenberg attacked Lavater's physiognomy for being an outburst of sentiment, penned in a language of initiates, fashionable but vague and spoken
only by self-identified geniuses.[105] As an alternative, Lichtenberg proposed
a pathognomy that interpreted the movements of the body as the expressions of specific frames of mind. His epistemological and educational project was brought to life in a series of illustrations by Chodowiecki titled
"Natürliche und affektierte Handlungen des Lebens," published in the
Göttinger Taschenkalender of 1779. It compares the postures of partici-
pants in a range of social situations that were likened to scenes on a stage:
"scenes from that play that we see performed every day, and in which we
not infrequently play a role."[106] Lichtenberg and Chodowiecki present such
fundamental rituals of contemporary society as the lesson, the conversation, the prayer, the promenade, the salutation, and the dance. In the case
of the promenade, a contrast between the "natural" and the "affected" version is achieved not only by the contrast between straight and crooked
lines but also by contrasting details of clothing and the social interactions
between promenaders and their companions. The natural promenade appears here as an educational opportunity, a chance to teach children not
only proper posture but also—in exercising *Zärtlichkeit* (tenderness) with
their parents as a role model—appropriate moral conduct (fig. 1.7).

Figure 1.7 "Der Spatziergang/La promenade." From Daniel Chodowiecki and Georg Christoph Lichtenberg, *Natürliche und affektierte Handlungen des Lebens* (1779).

In the seventeenth and eighteenth centuries, the semiotic understanding of promenading was still codified in a rather schematic form in conduct books and medical and educational treatises. That understanding became more nuanced, however, as new social spaces emerged in which the increasing intermingling of social classes allowed for observing more diverse forms of gait. The so-called promenades were semipublic walkways for casual strolling built in or on the outskirts of urban areas. Increasingly, they came to define not only the outer bounds of squares and gardens but also a specific manner of walk.[107]

Late-Enlightenment walkers found it quite natural to detect class status or individual traits in a person's gait: this is evidence of the same tacit or practical knowledge that informed the moral semiotics of Lavater and

Lichtenberg.[108] Because this embodied knowledge of walking was tied to a
concrete and public milieu, it is unsurprising that critical discourse on the
correction and direction of human locomotion was dominated by analo-
gies to both the theater and everyday life.

The theories of the popular philosopher Johann Jakob Engel on gait
and the language of gesture, published in his *Ideen zu einer Mimik* (1785–
1786), belong to this context. Lavater's writings received more attention
from later scholars, but in contemporary discussions of expressive movement, Engel's text was equally influential, and its influence stretched
far beyond the German-speaking world.[109] He did not intend it merely as
a theory of the dramatic arts but said that it would derive its "absolute
value" as part of a larger anthropological project, namely, the "Kenntnis
des Menschen" (knowledge of man's nature):

> We perceive the nature of the soul only through its effects, and we
> would learn more about it if we were to observe such effects—the
> many ways its various ideas and stirrings express themselves through
> the body—more diligently. As we cannot see the soul directly, we
> should look even more diligently and attentively at its mirror; or bet-
> ter, its veil, which is fine and flexible enough to allow us, through its
> slight folds, to divine the composition of the soul.[110]

Exactly like Lichtenberg's pathognomy, Engel's mimetics sought to
"observe how the body expresses the soul" by taking note of "fleeting
bodily movements" that could reveal a particular condition of the soul.[111]
But Engel also goes a step further in suggesting that idiosyncrasies of gait
could allow an attentive observer to become party to the flow of ideas
moving through the walker's head:

> Gait is the result of shadowy ideas that silently direct the will and
> whose succession is governed by those clear ideas that prevail in the
> moment: the former suffer from the influence of the latter; the latter
> from the influence of the former. Thus each specific mindset, each in-
> ternal stirring and passion, has its own distinctive gait; and that which
> Hercules's wife said of Lykus is equally true of every character: *Qualis
> animo est, talis incessu.*[112]

This argument extends the moral semiotics of human locomotion far beyond the works of Lavater and Lichtenberg to ultimately address every individual psychological impulse. Each step a person took was now under-

stood as participating in a kind of parallel progress in which every external movement had an internal counterpart.

The Enlightenment thinkers discussed in this chapter all strove to isolate the natural human gait from its artificial manifestations and to define and cultivate it by introducing new forms of description and measurement. In their efforts we see the rudiments of a true science of human locomotion emerging. In the accumulating body of knowledge on locomotion— which was in the Enlightenment largely founded on everyday observation of practices that had taken on a new cultural meaning, such as the stroll, the foot journey, and the soldier's march—both semiotic and mechanical approaches began to be articulated in various (and not mutually exclusive) registers. It was the military who hewed most resolutely to the mechanical model of walking as a series of prevented falls. The way military theorists defined natural gait overlapped with late eighteenth-century educators such as Vieth and Gutsmuths, who authored the first treatises on physical education. Both groups, moreover, focused predominantly on ideals of posture and movement oriented to the male body.

Notwithstanding intersections such as these, knowledge about walking in late-Enlightenment Europe, as examined here in a sampling of institutions and practices, can hardly be characterized as part of a unified discourse or the expression of a larger process of civilizing or disciplining bodies. Rather, "good" walking was something that resulted from obeying rules that derived their authority from the laws of mechanics while still remaining subject to both aesthetic and moral criteria:

A good gait requires the knees to straighten with every step, the feet to land opened a little outwards, the heel not to strike the ground too decidedly or too long before the ball of the foot, the foot to treads straight without placing more weight on the inner or the outer edge. . . . Moreover, the steps must be neither too long nor too short. They must be in a particular relationship to the overall height of the body and especially to the length of the lower limbs, a relationship that each person will sense for himself. Too long a stride betrays effort; too short a stride affectation. . . . The carriage must be upright; the head must not hang toward the ground as if searching the terrain. Some people also have the ridiculous habit of periodically looking down at their legs, apparently owing to some minor vanity regarding the elegance of their *chaussure*. The arms should normally follow the movement of the lower limbs so that right foot and left arm, and left foot and right arm, move together at the same time, or nearly so. This is natural and the result of the way

the body is built; still, the movement of the arms must not be too conspicuous. They may be held entirely still or allowed to move a little,
but to swing them, as if one wanted to throw them away, is unseemly.
Good gait further requires proceeding in one set, straight direction and
not wandering back and forth across the line like a drunkard.[113]

In the normative definition of "good gait," then, medical and scientific
arguments about natural, functional motion rooted in the constraints of
human anatomy were inseparably associated with the standpoint of moral
semiotics. Not until the nineteenth century would that association begin
to unwind.

CHAPTER TWO

Observers of Locomotion:
Theories of Walking in the French
Science de l'homme

In the late eighteenth century, knowledge about human locomotion was fragmented and heterogeneous. The first major attempts to integrate this fragmented understanding into a coherent scientific framework came at the turn of the nineteenth century in France. The project coalesced around a group of natural scientists and physicians who, in the years around 1800, set out to establish a new scientific discipline: what they called the *Science de l'homme*, the "science of man." In practice, they approached that goal by different routes, but they all shared a common discourse, one that insisted on placing the natural history and philosophical anthropology they had inherited from the eighteenth century on a more secure empirical footing through the use of observation.

The science of man had roots in the holistic anthropology formulated by the vitalist Montpellier school, most especially by its foremost representative, Paul-Joseph Barthez.[1] Preferring observation to experiment, advocating a strict antimetaphysical epistemology, and, centrally, aspiring to illuminate the multiple links in humans between the moral and the physical, the Montpellier physicians formulated a program whose influence can be traced within intellectual culture and the life sciences in France far into the nineteenth century.[2]

The diversity of approaches represented within the science of man was reflected in the membership of the new Société des observateurs de l'homme, founded in Paris in 1799. It counted among its most prominent members the *médecins-philosophes* Pierre-Jean-Georges Cabanis and Philippe Pinel, both strongly influenced by Montpellier vitalism.[3] That so many physicians in the years around 1800 came to regard the natural human gait, in particular, as a central problem for the science of man was largely owing to the cultural, economic, and political upheavals brought on by the French Revolution that had made Paris itself into a complex field of medical observation and experiment. These developments are commonly associated with the ascent of anatomical pathology in the great Parisian hospitals.[4] But although new interventionist methods were indeed on the rise, they did not replace the taking of detailed bedside observations of living patients. The goal of nosography, for instance—which developed in the eighteenth century out of the work of vitalist physicians ranging from François Boissier de Sauvages to Pinel—was to describe in words, as completely as possible, all the relevant symptoms of disease, including characteristic forms of gait and gesture.[5] The virtues of medical observation were similarly inarguable for those exponents of the *Science de l'homme* who engaged critically with classic texts on the mechanics of locomotion. And for most of the physicians of this period, who were inclined to vitalism, extending mechanical theories of locomotion remained a central goal. Giovanni Alfonso Borelli's *De motu animalium* thus remained the canonical reference within a rapidly growing body of literature on locomotion physiology, both human and animal, into the 1830s.

By around 1830, however, the vitalist discourse in locomotion physiology was beginning to run its course. This was partly owing to an internal conflict: this discourse prescribed the exhaustive empirical observation of human and animal locomotion, but in practice the observational field was staked out by just a handful of canonical texts. But partly, too, the breakdown can be ascribed to the parallel steady ascent of a new experimentalist physiology that relied on vivisection and the findings of anatomical pathology and sought to move the study of human and animal mechanics behind the closed doors of clinics and laboratories. Still, it is important to note that this early experimentalist physiology made no radical break with the holistic science of man. Rather, as we shall see, experimentalism won ground through a process of gradual distancing, provisionally retaining some of vitalism's set pieces although without ever integrating them fully. The limits of the vitalist medical discourse finally became clear with its annexation by literary works such as Balzac's *Comédie humaine*.

A VITALIST MECHANICS OF LIVING BEINGS

The famous Montpellier physiologist and physician Paul-Joseph Barthez (1734–1806) first announced the "elements of a new science" in his *Nouvelle mécanique des mouvements de l'homme et des animaux* (1798). The book presents the results of Barthez's twenty-five years of critical engagement with Borelli's *De motu animalium*—that magnum opus of one of the foremost exponents of iatromechanics (a school that sought mechanical explanations for physical and medical phenomena).[6] Ever since its posthumous publication in 1680–1681, Borelli's treatise had been the key scientific reference on its subject, but it had also found readers from much further afield.[7]

The article "Debout," for instance, on the upright human stance and gait, written by the physician Arnulphe d'Aumont for the fourth volume of Diderot and d'Alembert's *Encyclopédie* (1754), is largely based on *De motu animalium*, which it calls an "incomparable work."[8] D'Aumont's text is basically a résumé of the Borellian theory that humans and animals make forward progress by applying muscular force to counteract the force of gravity and propel the weight of the body across the ground. The iatromechanical school regarded living organisms as machines of muscle and bone that obeyed Galilean laws of nature.[9] As such, it was possible to extrapolate living movements from anatomical experiments and also from measurements taken from dead bodies. The method Borelli suggests for determining a body's center of gravity exemplifies this approach: in order to define that imaginary point at which the forces of gravity acting on the various parts of the body are in equilibrium, a cadaver stretched on a measured board was balanced atop a V-shaped block (fig 2.1, illustration 12).

Following Borelli, in the *Encyclopédie* d'Aumont describes upright posture and gait as an act of falling that is prevented by the work of muscles and mechanical counterforces: "For a man to stay on his feet, no matter the position of his body, whether leaning, curved, or bent, requires merely that a line drawn from the center of gravity—which according to Borelli . . . lies between the pubic bone and the buttocks—should end in the square space of ground that is occupied by the soles of the two feet and the space between them."[10] Should the line of gravity stray outside the virtual parallelogram outlined by the two feet when walking, "no effort of the muscles could save the man from falling, unless the weight of his body were counterbalanced with the aid of some mechanical force."[11] The falling man might counterbalance by swinging his body abruptly in the opposite direction, or taking advantage of air resistance by "seeking support, so to speak, in the air all around, and beating it, so that its resistance should repel the body back toward the center of gravity from which it had strayed."[12]

Barthez was both friend and physician to d'Alembert, and he himself became a collaborator on the *Encyclopédie* in 1754. As such, he was undoubtedly familiar with d'Aumont's article.[13] In the *Nouvelle mécanique*, however, he sought to relegate the mechanical perspective to what

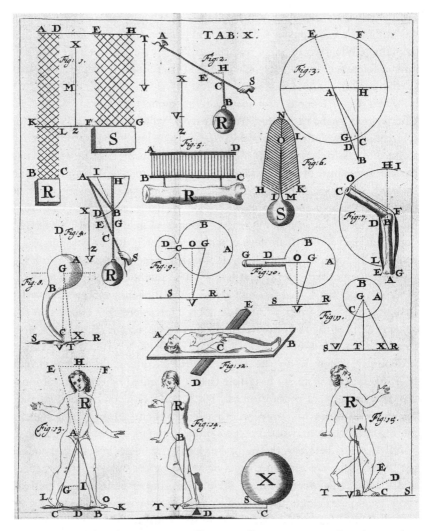

Figure 2.1 Table 10 from Giovanni Alfonso Borelli, *De motu animalium* (1680–81).

he considered its proper place within the physiology of locomotion. For
the Montpellier school, the engagement with Borelli and his influence
on eighteenth-century medicine was critical. Like Georg Ernst Stahl before them, the Montpellier vitalists did not think that organic processes
could be explained solely in terms of physics and chemistry, as the iatromechanical conceptualization of living beings assumed. But Barthez's famous doctrine of the *principe vital* (vital principle) was also a departure
from the basic tenets of Stahl's animism, which held that all the voluntary

and involuntary movements of the body were expressions of a metaphysical, spiritual entity.[14] Here, the Montpellier school saw themselves purely as observers of human nature, fact collectors in the Newtonian tradition. In theory, at least, they eschewed any hypothesizing about the interior nature of the phenomena they observed. The task they set themselves was simply to describe and classify physical process and events impartially.

Barthez was committed to this empirical epistemology; thus, he scrutinizes the historical reception of *De motu animalium* and the theories of mechanical physiologists for any trace of the metaphysical. He attacks authors (such as Albrecht von Haller) who, following Borelli, try to "explain the walking and jumping movements of humans and quadrupeds in terms of a *reaction against* or *repulsion from* the ground."[15] Here, Barthez is specifically targeting a theory of "imaginary repulsion," derived from a deductive approach to mechanics, that stated that "when the leg is extended, and pushes against the ground with the tip of the foot, the body is propelled forward by a reflected movement, resembling the movement of a barge that departs from the bank when the bowman pushes off with his stick."[16]

Barthez himself theorized that in "the ordinary gait, or the most natural," the legs acted on the pelvis in a series of individual impacts, by this means propelling it forward.[17] Hence, continued movement was neither solely the result of mechanical forces nor the result of willpower alone. Instead, Barthez thought, the action was distributed over multiple muscles, and according to his doctrine of the vital principle, each muscle possessed an individual force. He calls this the "force of fixed situation in the molecules of muscle fibers" (*force de situation fixe des molécules de leurs fibres*), a hitherto unrecognized "life force" (*force vivant de ces fibres*) that, in contrast to Haller's notion of irritability, was only ever postulated, never demonstrated in animal experiments.[18] Barthez invokes the "force of fixed situation" to resolve a famous problem raised by Borelli: namely, that muscles had to exert great force to lift or move small amounts of weight.

In a 1757 article for the *Encyclopédie*, "Force des animaux," Barthez had reviewed a number of physics experiments on the amount of work done by human and animal bodies that had led him to conclude that "animal forces" could not obey the same laws as inert bodies.[19] In the major work *Nouveaux éléments* (1778), he suggests a way to explain the great physical effort of which muscles are capable: "the action of the vital principle in the fibers of a muscle can, with a great degree of variability, set and hold the molecules in a fixed, close position such that this force of permanent situation of the molecules can overcome the efforts of the considerable forces that would tend to overpower them."[20] But the variable and unpredictable distribution of these forces, Barthez thought, could also be the cause of many pathological phenomena involving the involuntary speedup of the locomotor apparatus. As evidence, he cites the cases of a rare "disease in which one only can run, but not walk," described in Boissier de Sauvages's *Nosologie*.[21]

Vitalist physiology thus sought to dismiss Borelli's mechanical teachings on two key grounds: first, by arguing that the forces that caused motion in living organisms were principally unpredictable, and second, by citing pathologies of gait, saying that "natural" gait could not be defined except through the observation of such pathologies. Barthez's firmly held view of experiment (*expérience*) and observation (*observation*) as antagonists was rather uncommon for the second half of the eighteenth century and can certainly be seen as characteristic of French vitalism. The epistemological stance adopted by most scientists was that observation and experiment were inextricably intertwined.[22]

Before Barthez, another Montpellier physician and fellow *Encyclopédie* contributor also distanced himself from this widespread popular stance. Ménuret de Chambaud structured his articles on "Observation" and "Observateur (*Gram. Physiq. Méd.*)" as polemics in which the observer of physical phenomena, possessed of genius and vision, was pitted against the experimentalist. The observer, de Chambaud writes, "is modest enough to study phenomena as nature presents them to him." The experimentalist, meanwhile, "combines things himself, and sees only the results of his combinations."[23] De Chambaud's characterization of the observer, whose genius and keen insight allow him to see nature unveiled—to read it like a book—relies on the old literary topos of the legibility of nature.[24] In Barthez's writings, this topos was coupled with a scientific method that allowed for building theories based on observations drawn not only from the medical and scientific literature but also from the monuments and poetry of antiquity. As Barthez writes, "In the paintings they made of the movements of men and animals, the great poets, and especially Homer, who is surely the foremost among them, marked many traits that none of their commentators have well apprehended but that strike the experienced eye of a physiologist with a most singular impression of truth and inspire him to search for their causes."[25]

This put the poet in the position of a scientific observer avant la lettre. It also put the physician Barthez in dialogue with antiquarian and archaeological scholarly traditions,[26] as he constantly drew evidence for his arguments not only from clinical cases but also from excerpts of classical tragedies and epic verse as well as famous sculptures of antiquity. He hypothesized, for instance, that "the hands of the statue of Diana of Ephesus rest on spears" because the spears—like the headdresses extending to the shoulders of some Greek and Egyptian statues—helped keep the body in balance.[27] Although Barthez here rejects the search for any "mysterious" meaning in favor of a simple mechanical explanation, he also quotes repeatedly from works by Pliny, Plutarch, Strabo, Homer, Virgil, and Ovid to illustrate the incredible diversity of gait to which illness, age, and culture can give rise. Virtually none of his rather scanty, sparsely worded clinical observations end without seeking support from some classical author.

Looking at the *Science de l'homme* as it emerged in the early nineteenth century, it is clear that the name denoted neither any unified theoretical program nor any single methodology (its general emphasis on observation and analysis notwithstanding). Vitalists such as Barthez; materialists such as Cabanis; the philosopher Maine de Biran, who espoused a dualist spiritualism; utopian socialists such as Henri de Saint-Simon and his followers—such disparate thinkers as these all flew the new banner of the science of man.[28] And even within single theoretical camps, we can trace contrasting methods of observation and demonstration.

Barthez himself, near the end of his life, became involved in a number of fierce controversies and battles for primacy in the field with one of his former pupils, Charles-Louis Dumas (1765–1813), who was appointed to the anatomy and physiology chair at Montpellier in 1795. In his *Principes de physiologie, ou Introduction à la science expérimentale, philosophique et médicale de l'homme vivant* (1800), in which long passages from Barthez may be found (paraphrased and stripped of their classical references), Dumas attacks his former master's theory of jumping. Barthez had theorized that by contracting in opposing directions, the extensor muscles of the knee and hip generated a special force that could thrust the body upward. Dumas challenges Barthez's theory with an isolated and bizarre case from the anatomical cabinet in Montpellier: the skeleton of a professional vaulter that displayed a severe deformity. As an accompanying illustration showed, the vaulter lacked femurs, and thus knee joints. Dumas argues that the existence of this "singular case" seemingly refutes Barthez's theory. He asks, "Would it be possible for muscles acting on such short levers, and attached so close to the consecutive joints of spine and hips, to produce in the body a projective force sufficient to lift it from the ground and make it jump?" (fig. 2.2).[29]

Barthez declined to enter into a detailed anatomical discussion of the skeleton. He responded to Dumas—never mentioning him by name—by

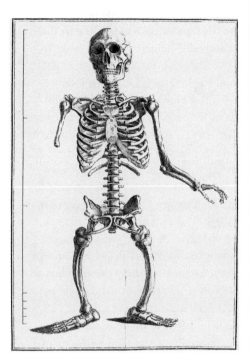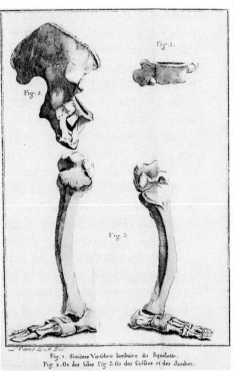

Figure 2.2 *Left*, Skeleton of a vaulter from the anatomical cabinet in Montpellier. *Right*, Detail of individual bones. From table 4 in Charles-Louis Dumas, *Principes de physiologie*, 2nd ed. (1806).

citing another, even more extraordinary case, not from his own observations but from Montaigne's famous essay *De la coustume* (Of custom). In the passage in question, Montaigne recalls as a child seeing a man who "flourished a two-handed sword, and . . . handled a halberd with the mere motions of his neck and shoulders for want of hands; tossed them into the air, and caught them again, darted a dagger," and performed many other such feats.[30] Having apparently misread Montaigne, Barthez says it was a child who possessed these extraordinary capabilities. He draws an analogy between this case and that of the Montpellier vaulter and suggests as a general theory that the extensors of the upper and lower spine, by contracting in opposing directions, might generate "extraordinary forces"—"in spite of the fact that the segments of the muscles that move each vertebral joint can extend only a very little, and act upon very short levers."[31] But this did not put the controversy to rest. In the second edition of his *Principes de physiologie*, Dumas countered that "one single fact taken from the writings of Montaigne, who surely was no great authority on this matter," was hardly sufficient to topple his detailed argument.[32] Besides, Dumas argues, the two observed cases were not truly analogous, for the weight of a body in the act of jumping was supported by the ground, which was not the case with objects moving through the air. Concerning human movement, Dumas ultimately reaches a skeptical conclusion that sought to restrict even further the validity claims of mechanics in the realm of physiology:

> Facts of this kind will always be difficult to explain. We struggle even to comprehend the prodigious jumps of which some people are capable; and we are equally embarrassed in our attempts to explain them. Indeed, it is quite impossible to reduce to a calculation the human functions that would seem to lend themselves most to it, for such functions are frequently altered and complicated by laws that apply only to living bodies, as well as by circumstances of custom, necessity, and education, which exert more power over human nature, perhaps, than does the mechanical organization [of the body].[33]

"TRANSCENDENT PHYSIOLOGY": A MEDICAL SEMIOTICS OF GAIT

In Paris, meanwhile, new approaches in physiology had already sprung up that took a different view of human and animal movement and would ultimately displace the teachings of Barthez.[34] In particular, the work of Xavier Bichat (1771–1802) attracted numerous followers and exegetes. Bichat introduced an experimentalist approach to vitalist physiology along with a new theory of life itself. He famously posited that animate beings had "two lives": the "animal" and the "organic." The "animal life" (*vie animale ou de relation*) included the locomotor and vocal apparatuses, whose functions were regulated by the brain and the conscious mind. The "organic life" (*vie organique*) proceeded outside of voluntary control. It included the digestion, the circulatory system, breathing, and glandular secretion.[35] This was a fundamentally dualist model that broke completely with the epistemological postulate of a single unified vital principle and rendered attempts like Barthez's to formulate a mechanics of animate beings untenable.

Bichat died young, leaving his work on anatomy and physiology unfinished. His ideas, however, were quickly integrated into psychological and anthropological theories of expressive movement and gesture. This may be clearly seen, for instance, in the work of Bichat's cousin, Matthieu Buisson (1776–1804), who published the final two volumes of Bichat's *Traité d'anatomie descriptive* and whose distinction between the "active" and "nutritive" lives is inspired by Bichat:

> It is locomotion considered as a function, or quite simply general locomotion, that specifically belongs to the active life: for most of the actions directed by the will are executed through this kind of locomotion; through it we move from one place to another; [through it] we act on external objects to make them serve our purposes; to it, essentially, the sense of touch belongs; finally, it is what the intelligence uses to express itself through gesture, which—just like speech—is a distinctively human faculty.[36]

Buisson's contention that gesture, broadly defined, was a purely human attribute represents a break with Bichat. According to Bichat, human intellect and judgment were expressed solely through the muscles of the face and neck, and only animals possessed an instinctual language of physical gesture.[37] Buisson's broader conception of gesture led him to conceive all the movements of the human body as a *fonction de relation*: a function of the relationship between organism and outside world. This meant that walking could not be understood separately from the senses of sight or touch: "the natural use of movement in the lower limbs is necessarily tied to an awareness of the objects all around us, which can delay this movement, impede it, or make it end in disaster."[38] For this new variation on vitalist locomotion physiology, which could read gesture not only in the motion of the hands but also in the "entire system of general locomotion,"[39] the model case was the gait of a blind person. It provided an *ex negativo* demonstration of the intimate connection between seeing and walking:

> A solid object might halt the feet as they change places with one another; perhaps a hole in the ground has opened up in the place one wants to go. Such are the fears of a blind man left to himself: or rather, being incapable of accurately detecting objects and empty spaces, he has only a sufficient idea of the restricted space that he actually occupies, and he will not wish to leave it, just as we all are disinclined to walk in dark places.[40]

Predictably perhaps, Buisson rounds out his argument by citing the gestures of deaf persons, whose eloquence was matched by a heightened visual acuity. Physiological discussions such as these of sensory function in the blind and deaf had a famous philosophical precedent in the letters of Diderot as well as a cultural reference point in the well-known public events held by the Institut national des sourds et muets and the Institut royal des jeunes aveugles in Paris (schools for the deaf and blind respectively).[41] For Buisson, blind and deaf persons were stereotyped figures who invited psychological speculation. Here, the physiology of locomotion entered a moral realm, a world seemingly divorced from mathematical calculations and mechanical laws.[42]

In a succession of vitalist physicians after Bichat, we can observe an increasing decoupling of the anthropological observation of gesture from the anatomical and physiological study of the locomotor system. A semiotics of gait began to detach from the mechanics of walking, and the latter now receded into the background. This trend is evident in the entries on "locomotion," "marche," "motilité," "movement," and "geste" written by the anatomist and surgeon Pierre Joseph Rullier (d. 1837) for the *Dictionnaire des sciences médicales*, published between 1812 and 1822.[43] In this new medical semiotics, attention to specific gaits and gestures took on a crucial role as it was thought that involuntary bodily movements could lead a trained observer to a correct medical diagnosis. For while locomotor movements were understood to be purposeful and controlled by the will, physical gestures and facial expressions were, for the most part, not considered subject to conscious reflection.[44]

The assessment of normal and pathological sequences of motion took place against the backdrop of a system that classified diseases and disorders based on how much control ailing subjects exhibited over the *force motrice*, or "motive force." Rullier differentiated between disorders that weakened or eliminated the motive force—including apoplexy, lethargy, paralysis, paraplegia, and hemiplegia—and those—such as mania and other forms of insanity—that led to its "ruinous stimulation." In the latter type of illness, the "locomotor phenomena" were "exaggerated and multiplied in an unbelievable fashion, subject to a bizarre and fluctuating will whose acts bear constant witness to a disordered understanding of the mind."[45] Thus, for clinics that treated nervous and mental disorders, keeping detailed records of the numerous and characteristic disruptions of the locomotor system was imperative:

> In cases of *dementia*, who is not struck by the link between the confusion of gestures and the swift unbroken sequence of isolated, incoherent ideas, or the disparate emotions that these patients experience

without cessation? What inconstancy, indeed! What perpetual variability in the expression of the physiognomy! It offers a moving tableau
of images, each following rapidly on the last before it, too, is wiped
away; the lines of each are poorly drawn; they flow together and none
leaves a lasting impression. It is known, moreover, that the insane continuously shift place, position, attitude; and most often the disorder
and singularity of their *gestures* alone suffice to convey the confusion
of their ideas.[46]

A physician seeking to interpret the "disorder and singularity" of this
"moving tableau" could no longer merely rely on the repertoire of classical
rhetoric, although the classics still served as an obligatory point of reference for *aliénistes* such as Philippe Pinel and his pupil Étienne Esquirol.[47]
The symptoms of mania, especially, which were a subject of keen interest
for Pinel, presented a challenge for the psychiatric observer. Manic attacks
were accompanied by "vigorous, brusque, and insecure" movements and
seemingly absurd and "ridiculous" gestures: "The maniac is a Proteus;
hiding in every possible shape and form, he eludes observation by even
the most trained and attentive eye."[48] Faced with the epistemological un-
certainty that necessarily accompanied such an elusive object as insanity, clinical semiotics relied on the assumption that a patient's perceptible
gestures, gait, and words were direct expressions of an impenetrable inner
reality.[49]

Alongside this medical ethos of pure observation, as it sought to divest itself of the framework of classical rhetoric, the physiognomy of
Lavater also enjoyed a revival in France. A ten-volume French translation
of Lavater's *Physiognomische Fragmente* was published between 1806 and
1809. Its subtitle alone clearly demonstrates how far removed it was from
the German original: *L'art de connaître les hommes par la physionomie,
par Gaspard Lavater. Nouvelle edition, corrigée et disposée dans un ordre plus méthodique* [. . .] *; augmentée d'une exposition des recherches
ou des opinions de la Chambre, de Porta, de Camper, de Gall, sur la
physionomie.*

The driving force behind this ambitious project was the *médecin-
philosophe* and "littérateur distingué" Louis-Jacques Moreau de la Sarthe
(1771–1826). Moreau was for many years a librarian at the École de médecine in Paris, and he also belonged to the Société des observateurs de
l'homme.[50] His revised edition of Lavater, along with a series of popular
scientific lectures at the Athénée de Paris, launched a reception tradition
that would have far-reaching effect as it tried to strip Lavater's work of its

rapturous, theological attributes and reshape physiognomy into a rigorous observational science of expressive human gesture.[51]

For the scientific critique to which physiognomy had constantly been subjected, Moreau mainly blamed the fragmentary and unsystematic nature of Lavater's work. He called Lavater "neither a physiologist nor a physician; in fact not a natural scientist at all."[52] In his translation of the *Physiognomische Fragmente* (made with the anatomist Jean-Joseph Sue fils), Moreau abandons Lavater's original structure and rearranges the text according to a completely new system, dividing it into thirteen "Études de la physionomie" that, in terms of analytical observation, move from the simple to the more complex. Moreau's reading of Lavater's physiognomy was informed by the work of *idéologues* such as Cabanis and Destutt de Tracy. In Lavater he found a new, "transcendent physiology"—"an order of thought that is highly elevated, in which diffident and attentive observation, free and bold imagination, the utilitarian research of the scholar, and the sublime meditations of the philosopher, converging and nearly conflating, illuminate and augment one another through mutual communication."[53] This new "genre des connaissances mixtes" was situated "in the tableau of human knowledge between the moral and the physical sciences, filling the gap between them."[54]

Although Moreau claims to have retained Lavater's text in its entirety, it is perhaps more accurate to say that he dissolved it, dispersing its elements within a new body of text that was expanded with numerous editorial excursuses and glosses, texts quoted and paraphrased from other authors, and some six hundred mostly new illustrations. These additions were meant to situate physiognomy within the framework of comparative anatomy and physiology. They were also intended to extend its empirical reach to include the multitude of pathognomic phenomena discussed by Engel in his *Ideen zu einer Mimik*. Long stretches of the book consist of passages or paraphrases and illustrations from Engel inserted by Moreau to supplement Lavater's brief treatment of posture and gait.[55]

Like Engel and Lichtenberg before him, Moreau wrote not only for painters, sculptors, and actors but also for a general audience of educated readers to whom he offered an empirical method for achieving personal insight. "I maintain," he writes, "that the accurate depiction of a dozen or so of our body postures, selected intelligently from those moments when we believe ourselves alone and unobserved, can lead us to greater self-knowledge and become a source of useful instruction. Perhaps no more than this is needed to give a complete idea of the character of every individual."[56]

There were countless individual variations in gait, and different gaits and postures were characteristic of different professions and classes. Moreau believed these variations represented a mixture of natural forms and ones based on imitation. But he also considered even the mimetic elements of gait and the mannerisms they produced as, ultimately, "results of Nature" that reinforced the "primal character."[57] This made gait one of the most significant markers in the project to develop a complete anthropological record of the "outer life" (vie extérieure) of humankind, including every "outward habit" from speech to dress and even interior design.[58]

In Moreau's commentary on the illustrations from Engel's *Mimik* and Lavater's *Fragmente*, the individual traits of the figures depicted are not his focus. Rather, he makes a summary assessment of their intellectual and moral virtues and defects. As an example, he contrasts a variety of postures typical of the "thinker" with gaits he classifies as "idiotic" or "feeble-minded" (fig. 2.3). In a series of twelve male figures assembled from Engel's book and placed in a new order, Moreau detects various perverted forms of correct observation and reflection—shown, for instance, in the deliberations of the sophisticate, which are governed by "cunning" and a "calculating spirit" or simply in "feeble," "incurious" minds. Moreau broke the tight connection Engel had forged between the iconic representation of gesture and the discursive problematization of its "translation." His approach was symptomatic of the enterprise of "new" physiognomy as a whole. Engel, a figure of the German Enlightenment, could contrast two well-known Italian gestures (warning of a devious person and dismissing a warning; fig. 2.4, illustrations 8 and 7, respectively) and discuss in detail the difficulties involved in describing and explaining them. Moreau, the French physician, perceived in the very same figures, "according to my own manner of seeing and feeling," the "deliberations of a man not made for reflection" and the "feigned indifference of a self-complacent man."[59] In place of a discursive reflection on the epistemological dilemma of the observer, who must figure out how meaning can be deduced from the gestures portrayed, he offers a facile classification of observers into good and poor—thinking and unthinking—via simple snapshots of communicative movements.

Moreau's practice of shifting and skewing these figurative representations of gesture and gait, which were linked to concrete cultural fields of meaning, was informed by a hierarchically organized typology of societal occupations that put intellectual labor at the very top. He writes that "every métier, every occupation should generally be seen as an ongoing, specialized, and lifelong education that develops, exercises, and fortifies

Pl. 18.

Figure 2.3 Plate redrawn from Daniel Chodowiecki. In Moreau
de la Sarthe, ed., *L'art de connaître les hommes.*

Figure 2.4 Plate redrawn from Johann Jakob Engel, *Ideen zu einer Mimik.*
In Moreau de la Sarthe, ed., *L'art de connaître les hommes.*

certain of the organs and establishes a particular relationship between the individual and nature."[60] Drawing on the physiology of Bichat, Moreau proceeded on the assumption that society as a whole could be divided into various distinct types, each of which was permanently defined by the predominant use of specific body parts.

Thus, the habitual imprints left on the body by working conditions (both physical and social) ultimately rose to the rank of privileged signs in the physiognomic portrait of society. Every class had "its own character and manner of expression: the craftsman, the nobleman, the commoner, the man of letters, the clergyman, the magistrate, and the soldier."[61] For Moreau, then, the "true basis of human inequality" was the "diversity of uses" to which the parts of the body were subjected in different specialized activities. At the top of this hierarchy stood occupations dominated by the "intellectual organ"; in the middle were the arts; and at the bottom were the purely manual jobs, "all of whose activities belong to the animal life."[62] Thus physiognomy, recast as physiology, with an empirical foundation in the observation of characteristic manners of gesture and gait, now presented itself as a totalizing discourse, a "physical and moral anthropology."[63]

ANIMAL MECHANICS, DISSECTED

In 1809, a young anatomy prosector named François Magendie (1783–1855) made a call for finally closing the cover of what he called the "novel of physiology." "The majority of physiological facts," Magendie proposed, "must be verified by new experiments. . . . This is the only means of delivering the physics of living bodies from its present state of imperfection."[64] The new experimentalism he advocated would investigate how life worked in fact, independently of the generalizing anthropological theories of the "science of man." It set physiology firmly on a new course that would ultimately lead to its autonomy, both institutional and scientific, from comparative anatomy, natural history, and philosophical anthropology.[65] Magendie's *Précis élémentaire de physiologie* was the earliest handbook in experimental physiology. First published in 1816–1817, by 1838, it had been republished five times and translated into multiple languages.[66] Yet despite Magendie's programmatic rejection of vitalist medicine, what can be observed in practice is not so much a radical break with the "science of man" as a process of gradual disengagement.

Magendie's initial statements on the physiology of locomotion seem eclectic and disparate. They relied not on experiment but on his reading or at best on self-observation. Many aspects of his thinking echoed the vitalist physiology of Barthez and Bichat: for instance, his consideration of the links between walking and the senses (especially sight), or his psychological discussion of gesture and the bodily expression of the passions.[67] Although Magendie was convinced of the need for a new, "complete treatise on animal mechanics" that would go beyond Borelli and Barthez, he first had to pay tribute to both authors. He praised (and adopted) Barthez's theory of jumping while at the same time describing human skeletal mechanics in a way that accorded perfectly with Borellian iatromechanical principles.[68]

Magendie entrusted the actual composition of the new treatise he had called for to one of his first collaborators: the young, newly graduated physician François-Désiré Roulin (1796–1874),[69] who led off a new article series in the *Journal de Physiologie Expérimentale* (founded by Magendie in 1821) with a critical review of previous literature on locomotion. The main aim of Roulin's article was to undermine the authority of Barthez's *Nouvelle mécanique* and restore the iatromechanics of Borelli to its rightful place. Roulin did not discuss Barthez's study in much detail, seeing in it nothing more than "misplaced erudition." He asks, "How relevant, indeed, is an author who begins a treatise on animal mechanics by pontificating on the utility of oriental languages? Why do we need to know the Arabic name for the walk of a stumbling woman or a sauntering donkey? What use have we for his hypotheses on the purpose of Harpochrates' plaits, or the spears of Diana of Ephesus?"[70]

By comparison with the experimental method favored by Magendie, which involved not only observation of the movements of living organisms but also intervention in the form of vivisection, the research conducted by Barthez, with its borrowings from classical authors, appeared antiquated indeed. Roulin scoffs, "He has consulted animals far less than books and believed his erudition could take the place of observation and experiment."[71] Such statements represented a programmatic rejection of the holistic anthropology of the science of man. Rather than taking the varied cultural and pathological manifestations of human and animal gait as his starting point, Roulin advocates a rigorous return to the mechanical laws so exemplarily demonstrated in Borelli's *De motu animalium*— "numerous errors" and a certain "lack of clarity" notwithstanding.[72] In the same year in which he published this first article, however, Roulin left Paris to take up a professorship in physiology in the republic of Gran Colombia (recently founded by Simon Bolivar). For a time, the vision of a Leibnizian, mechanical physiology of locomotion, where the human organism was "a kind of automaton, animated by a double wheelworks," remained a plan and nothing more.[73]

For the second edition of his *Précis* (1825), Magendie made fundamental changes to the chapter on locomotion. He wrote a new introduction in which he repeated some of the maxims from Roulin's article verbatim. He also inserted a long section addressing the influence of the brain on locomotive activity.[74] This became a central passage in the revised textbook, which presented "new facts" based on Magendie's controversial laboratory experiments in which he carefully observed and documented the movements of animals after destroying specific parts of their brains or spinal cords.

Magendie demonstrated, for instance, that removing the corpus striatum from the brains of living rabbits caused the rabbits to lose control of their movements. As he notes, "the animal immediately leaps forward and runs very fast; if it stops, it remains poised for flight. This phenomenon is especially striking in young rabbits; it almost seems that the animal is propelled forward by an internal force that it cannot resist; running at speed, it will sometimes leap over obstacles in its path, but it does not see them."[75] Other animal experiments led Magendie to posit the existence of four "spontaneous impulses" or "forces" in the brain that moved the animals forward, backward, or sideways, like automatons.[76]

As recent historians of physiology have stressed, many members of the newly founded Académie de médecine in Paris regarded Magendie's animal experiments (which had received their initial support from the Académie de sciences) as model examples of a new experimental methodology.[77] Even so, in order for the new experimentalism to gain prestige among physicians, it was crucial to demonstrate a close reciprocal relationship between laboratory and clinic. Thus, Magendie and his staff constantly sought to point out connections between their animal experiments and the unexplained pathologies of gait that they had observed in multiple patients and that had already been described at length in the nosographies of the *aliénistes*. Magendie worked with clinicians to gain access to large Paris hospitals, including the Pitié and the Salpêtrière, where Pinel and Esquirol practiced; in return, the clinicians received the chance to publish in his journal.[78]

Beginning with the second edition of the *Précis*, this institutional and practical alliance is recorded in the way that Magendie relates his animal experiments to many case histories of rare locomotor disorders. He describes, for instance, the case of a young man "of agreeable appearance [and] a cultivated mind, gentle and easygoing, but of great nervous susceptibility," who after his marriage experienced "intense grief" in the wake of "a mental illness that struck his wife upon the birth of their first child. Throughout her entire illness, the man did not leave his wife's side for an instant; he accompanied her on a journey, and so witnessed, over the course of nearly a year, the ravings and convulsive movements of one to whom he was most tenderly attached."[79] When his wife recovered, however, the young man was plunged into a "veritable melancholy," which was treated unsuccessfully with spa therapy. Eventually losing "all voluntary control over his movements . . . the unfortunate young man was thus compelled to move in the most irregular way, adopt the most bizarre postures, and make the most extraordinary contortions. It is impossible to depict in words the multitude and *strangeness* of his movements and poses. Had he lived in less enlightened times, people would surely have said he was possessed, for his contortions were so unlike any natural human movements that they might readily have been seen as the devil's work."[80]

In another case, a patient felt a constant impulse to walk: "Just when his apathy seemed greatest, he suddenly rose, began to walk in an agitated fashion, made several circuits of the room and did not stop before he was quite exhausted. One day the room seemed not enough for him, [and] he went out and walked for as long as his strength permitted . . . The following day he left again; his wife tried to prevent him; he became angry and would have struck her; so she let him go but followed him; nothing she could say to discover where he was going, or to make him rest, was of any use; only after he had walked for an hour and a half, aimlessly and as if driven by a *force he was unable to overcome*, did he feel fatigue and come to a halt."[81] Was this mysterious, insuperable force identical with the automatic stimulus Magendie had observed in his test rabbit? For Magendie, the moment of truth came at autopsy: "When the body was opened, several tubercles were found, involving particularly the anterior portion of the hemispheres. This makes it extremely probable that there exists in mammals and humans a constant force or impulsion that would propel them forward. In the healthy state, it is under voluntary control, and seemingly counterbalanced by another force acting in the opposite direction."[82]

Coupling a terse experimental protocol that involved the manipulation of animal mechanics by scalpel to a detailed case history from a mental clinic serves a key epistemological function in Magendie's handbook. The aim is to prove that experimentally induced phenomena in animals were transferable to pathological locomotor disorders in humans. The case history is even narrated in such as a way as to guide the reader's attention to the particular detail most essential to Magendie's argument, that is, the apparent influence of an unknown and involuntary force. But although

Magendie stresses the peculiarity of the melancholic's movements, he does not describe them in any concrete detail. Here language seems to fail him. This is the point where his physiology stops. He does not himself describe the pathological phenomena in any detail but is content to refer readers to the iconography of demonic possession.[83]

As such "extraclinical" references show, the association of experimental protocols and case histories also served to legitimize the experimental brand of pathological medicine Magendie was advancing—which may help explain why he included so many details that were ultimately extraneous to his argument. His pathophysiology promised to find causes for the many mysterious locomotor disorders recorded in the psychiatric nosographies of Pinel and Esquirol. But it also offered a justification for animal experiments, which were widely criticized as unnecessary and cruel, by fanning hopes for new therapies to replace the subtle, at times outré clinical legerdemain of the *traitement moral* (moral treatment) with which psychiatrists sought to influence their patients' imaginative faculties.[84]

THE "ADMIRABLE CALCULATION OF THE INSTINCT"

Not every physician of the Paris school was convinced, however, that physiology had to embrace animal experimentation. In 1830, Pierre-Nicolas Gerdy (1797–1856) expressed scathing criticism of the state of his discipline in a manifesto-like preface to his *Physiologie médicale, didactique et critique*:

> Their research laboratory resembles not so much a scholar's study as a place of murder and carnage where the laments and cries of animals expiring under dreadful torments ring out without cease. But Physiology is deaf to these creatures' pain; laughing, she cheerfully contemplates the horror of their movements and the anguish of their sufferings; and then, in an ecstasy over difficulties and obstacles surmounted, and comparing particularly the efforts and bodily fatigue of the experimentalist with the work of a great thinker in his study, she unhesitatingly places the manual labor of a man of experiment above the intellectual operations of a man of reflection. And there Physiology is consistent with herself: for the more active she is with her hands, the less rational her spirit becomes.[85]

Gerdy, who was appointed professor of external pathology at the Faculté de médecine of Paris in 1833, had previously expressed skepticism, if less harshly, about the experimentalist direction taken by Magendie and his students. As a young physician, in his first published work in 1821, he called physiology the "daughter of observation and reason." He contrasted the "simple contemplation" of naturally occurring phenomena with the numerous difficulties involved in correctly interpreting the "ambiguous language of experiments."[86]

Gerdy's writings exhibit a return to the epistemological positions taken by the vitalists in their confrontation with the mechanics of living bodies around 1800. But Gerdy had the firm ambition to surpass the work of Barthez, Dumas, and Bichat. Accordingly, in his *Physiologie médicale, didactique et critique*, whose first part was published in stages beginning in 1830—and which he submitted in application for a vacant chair in physiology at the Faculté de médecine in 1831 (he was initially passed over for his contemporary, Pierre-Honoré Bérard)—Gerdy made the analysis of human gait a central concern, calling it "one of the most complicated phenomena in animal mechanics."[87]

Gerdy published his initial work on gait in Magendie's journal in 1829. Doubtless he chose that venue because, like Roulin a few years before, he was sharply critical of the vitalist mechanics of Barthez and Dumas and advocated a return to the classical theory of Borelli. At the same time, he sought to reintegrate an anatomical and physiological investigation of human locomotion into the science of man as it had been conceived by Buffon. In this context, it comes as no surprise that Gerdy offered yet another elaborate rejection of the perception that "man, who walks always with his head lifted to the sky, treading the ground with his feet, owes this majestic posture to civilization alone," but "when living in the wild [he] goes on all fours, like the quadrupeds."[88]

Gerdy's texts on anatomy and physiology devote much space to the question of upright human posture and gait along with the question of racial diversity in humans considered from both a historical and a geographical point of view. By elucidating the multiple interrelationships between the physical and the moral realms as fully as possible, his writings served an ambitious plan: to synthesize the viewpoints of the individual arts and sciences, each incomplete in itself, and so transcend them. As he wrote, "The physiologist finds enlightenment in the works of theologians, moralists, writers in every genre, artists, and physicians; and he reflects back onto all their observations the light from his own in order to show man to himself just as he is, physically and morally, in nature and in society."[89]

In the years around 1830, physiology as a discipline was still broadly construed, and Gerdy's confident self-determination was very much in that spirit. It challenged the constriction of physiology to the laboratory-bound, experimental discipline envisioned by Magendie. Accordingly, Gerdy did not address the experiments on animal subjects that had been carried out in the interim, as he had principal doubts as to whether such work could really provide unequivocal data on the neurocerebral control of human locomotion.[90] Gerdy's physiology of locomotion was based on a close reading of Borelli's *De motu animalium;* it focused on anatomy and morphology and insisted on the careful observation and systematic ordering of details. It was as ambitious in methodological design as it was in scope. In the best tradition of the late eighteenth-century nosographies, Gerdy outlined a classificatory system for a "universal method" of observation. His aspiration was nothing less than to "perceive everything that can be perceived, and to record everything that can be recorded."[91]

Gerdy's physiology also drew heavily, though silently, on the simple, empirical, everyday observations of another Paris physician, Nicolas Philibert Adelon (1782–1862). On the streets of Paris, Adelon had found "a thousand different ways of walking." As he notes, "one need merely observe the various people in the street to recognize that perhaps no two of them walk in just the same way."[92] His eclectic *Physiologie de l'homme* (1823) devoted 236 pages to the topic of "locomotility, or the function of voluntary movement." It addressed every conceivable deviation from the act of normal forward walking: walking sideways, backward, on the knees, on tiptoe, on crutches, on a rope, on the hands, and last but not least, on all fours.

Gerdy's treatment of this subject, in which he also coined the word *musculation,* displays an equal fascination with anomalies of gait as opposed to "usual" upright walking. He even went down on all fours himself—"sans façons"—using his own body as a tool to establish just how uncomfortable that mode of locomotion was. He found "that it does not require much practice to walk just as fast on four legs as on two, or to run much faster than we usually walk on two legs."[93] He also addressed sideways walking, sidestepping sudden obstacles in the street, and how to avoid a "faux pas" when walking up the stairs, all based on his own self-experimentation.

Ultimately, however, the central and still unsolved problem for the French vitalist physiology of gait remained the true nature of normal, controlled human walking, the kind of walking that at each step breaks our fall and elevates humans above both the ground and the animal world. Adelon took it as a settled fact that natural walking, although it might seem automatic, was controlled by the will: "It is the will that sets the measure for how much the numerous muscles contract, which make us walk, and often this measure must be quite strict. At the same time, all these movements are executed with marvelous speed, and with practice they become so easy that they seem to execute themselves, and one can see in them no trace of voluntary control."[94] Gerdy's work also does not explain the regularity and the apparent automatism of normal gait through mechanics alone. Instead, he appeals partly to voluntary control and practice but partly also to a "calcul admirable de l'instinct" that unconsciously governs the mysterious vital force:[95]

> This calculation is quite admirable, for—having no means of measuring the force of the action of our muscles other than an obscure feeling, which discloses these actions and their energy to us secretly; and without thinking about it in the least—we send an impulse, on the one hand to our body weight (or, if you will, our center of gravity), and on the other hand to our lower limbs, and this impulse is in proportion to their own resistance and to the path they must travel to move forward such that our line of gravity leaves the base of support provided by the rear foot at precisely the same moment that the front foot is placed onto the ground.[96]

PHYSIOLOGY ON THE BOULEVARD: BALZAC'S *THÉORIE DE LA DÉMARCHE*

By the early nineteenth century, the mechanical and physiological project of the French "science of man" was shot through with internal tensions and contradictions. These were depicted with the greatest eloquence and originality not by any of the scientists themselves but by an outsider: the acclaimed author Honoré de Balzac. In 1833, at age thirty-four, Balzac published his *Théorie de la démarche* (Theory of walking) in four parts in the periodical *L'Europe littéraire*,[97] presenting himself as a competitor with the Paris physicians who recognized the everyday, more or less automatic movements of the human body as one of their most difficult problems.

That Balzac had scholarly ambitions is well known, and his connections to a number of scientists and physicians are well documented. He exchanged ideas with natural scientists such as Geoffroy Saint-Hilaire and the astronomer Félix Savary, a professor at the École polytechnique and librarian at the Bureau des longitudes who was interested in Borelli's iatromechanical theories. He also had a long-standing correspondence with the physician Jean-Baptiste Nacquart (1780–1854), who authored the *Traité sur la nouvelle physiologie du cerveau* (1809) as well as a number of articles for the *Dictionnaire des sciences médicales*. The *Dictionnaire*, in particular, has repeatedly been identified as an important source of both terminology and theory for Balzac.[98]

Given these starting points, it remains difficult to offer a clear picture of the significance of physiology within Balzac's literary works and its repercussions on the scientific culture of his time.[99] In 1834, he decided to divide his ambitious plan for *La comédie humaine* (*The Human Comedy*) into three groups of texts beginning with "Studies in Nineteenth-Century Mores," and moving on through "Philosophical Studies" to "Analytical Studies." The *Théorie de la démarche* was one of the analytical studies, alongside the *Physiologie du marriage*, initially published anonymously in 1829, and the *Traité de la vie élégante*, composed shortly thereafter. The latter works both took the form of treatises with a satirical slant. In several of the novels in the group of philosophical studies, problems of physiology figure as fatal traps that bring a series of fictional scientists to failure or the brink of madness. *Louis Lambert* (published in several versions between 1832 and 1836) sketches the portrait of a theoretician somewhat in the nature of a psychiatric case history; the theoretician admits defeat when faced with the thorny question of the nature of the will.[100] In *La peau de chagrin* (1831), Balzac summons contemporary representatives of Parisian science to stop the steady shrinkage of a magical ass's skin that holds the life force of the protagonist, Raphaël de Valentin. All their efforts fail miserably.

In an introduction (probably mostly written by Balzac himself) to the first edition of the *Études philosophiques*, editor Félix Davin loudly praises the scientific character of the project: "M. de Balzac—the very same!—writes *Louis Lambert*! He offers proofs, as the scholars do! . . . What the critics have failed to see is that *La peau de chagrin* is a physiological judgment, definitive, passed by modern science on human life; and that this work is his poetic expression of that judgment, abstracted from social individualities. The effect of desire, of passion, on the capital of human strength—is it not exposed magnificently here?"[101]

In view of the totalizing discourse conducted by Paris clinicians with the help of the physiological *Science de l'homme*, Balzac's literary annexation of their scientific credentials is not surprising. The boundaries of this physiology had shown themselves to be porous, as the discipline intermixed detailed observations of man's physical nature with psychological, historical, and cultural knowledge. And as the origins and early reception of Gerdy's *Physiologie médicale* clearly showed, attempts at a grand synthesis were fragile. The standalone preface to that work, published in 1830, included, in addition to a manifesto and an announcement of preliminary results, a plan for a four-volume work. But it took a year before the first half of the first volume appeared, and the volume in full was not published until 1833. Meanwhile, the remaining volumes, intended to address morals and psychology (including Gerdy's "théorie de l'entendement humain" [theory of human understanding]), were never published at all.[102] In one early review, the physician Frédéric Dubois d'Amiens derides Gerdy's long-winded program and additionally castigates him for his stylistic incoherence, his inept metaphors and neologisms, and his occasional citing of dubious historical anecdotes.[103] D'Amiens's scathing criticism predominantly addresses language and style, indicating that an unprepared reader would scarcely be able to tell whether a text of this kind was a scientific study or a "physiological sketch" from the pen of a literary man.

Although the satirical genre of the physiological sketch did not reach its full flowering in Paris until after 1840, when such sketches were disseminated on a massive scale throughout the city (a process Walter Benjamin interpreted in the light of his famous theory of the *flâneur* in the Arcades Project[104]), this late blossoming was merely the final stage in a widening of the semantic field that had begun as early as 1800. That field encompassed a broad spectrum, ranging from utopian theories of social and technological progress introduced by historians of philosophy in the Saint-Simonist movement under the catchphrase "social physiology"[105] to the specifically French anatomical and physiological reception of Lavater by Moreau de la Sarthe already discussed above to the bestseller *Physiologie du goût* (1825) by the judge Jean Anthèlme Brillat-Savarin, published in multiple editions. Brillat-Savarin, who portrayed himself to readers as a "médecin-amateur" in conversation with the leading vitalist physiologists of the day,[106] adopted a literary style peppered with neologisms and foreign words to sketch a "histoire morale" of taste and proclaimed gastronomy to be its own science with its own set of laws.[107]

The adoption of the totalizing discourse of physiology, visible in the mercurial style of Brillat-Savarin that bounced between axiom and anecdote, was continued by Balzac in his own inventive way. Above all, physiological discourse helped shape the character of a number of short essays that Balzac intended to collect into a "pathology of social life" for the group of analytical studies at the end of *The Human Comedy*. At the center of this project—which was never completed—stood the *Théorie de la démarche*, conceived in 1832 and employing an unusual form that has often been read as deliberately enigmatic.[108]

On its face, the *Théorie de la démarche* is a satire of a scientific treatise. It begins with a foreword and then turns to discussing a series of twelve dogma-like aphorisms, a "code of gaits" proffered for the use of France's new elite under Louis-Philippe, the Citizen King, "axioms to set feeble or lazy minds at ease, to relieve them of the pain of reflecting, and to lead them, by observing a few clear principles, to regulate their movement."[109] The colorful gallery of examples that Balzac provides, drawn from historical anecdotes as well as his own observations on the Boulevard de Gand, near the Paris Bourse (fig. 2.5), serve him mainly as illustrations of an axiomatics that derive largely from his reading of Lavater's physiognomic theories. For Balzac, gait is a "physiognomy of the body" or "thought in action,"[110] and to the attentive observer, peculiarities of gait mercilessly reveal both virtues and vices, industry and illness. On the boulevard, new members of the peerage betray their true social origins, while "well-corseted women" sway their hips like steam engines, revealing the "detestable precision" of their love lives.[111] The gaits

Figure 2.5 Claude-Louis Desrais and Etienne Claude Voysard, *Petit Coblentz ou promenade du boulevard des Italiens*. Musée Carnavalet, Paris.

of workmen, "condemned to repeat the same movement over and over,"
are contrasted with the bowed heads of students.[112] As discussed above,
Moreau de la Sarthe's French translation of Lavater—Balzac owned the
new 1820 edition[113]—outlined a similar physiognomic physiology of gait
and also classified social groups according to the "vie extérieure." Indeed,
Balzac uses many of Moreau's formulations almost verbatim in the *Théorie de la démarche*.[114] Moreau's hierarchy of manual and mental labor,
which can be traced back to Bichat, is taken up as well. Thus Balzac can
consider it "proven, that the marble-cutter was not born stupid, but rather
cutting marble for a living has made him so. His life is spent in moving
his arms, just as the poet's is spent in moving his brain."[115] Similarly, Balzac's procedure of watching Paris pedestrians from a chair on the boulevard must be understood within the tradition of anthropological observation espoused by the *Science de l'homme* in its quest to lay the empirical
groundwork for a sign language of gesture.[116]

If we read the *Théorie de la démarche* and its physiological analysis
of gait as only a preliminary to a larger goal, namely, the formulation of a
semiotics of urban society, then Balzac's many borrowings from the scientific and medical literature appear to be no more than "metaphorical
resources" that he deploys to analyze social reality.[117] Put negatively, they
become ideological props recruited to naturalize social inequalities.[118] But
although Balzac's brief treatise might look like a conventional satirical
social tableau relying on a shared, local cultural frame of reference (such
as the popular caricatures of Henri Monnier, the marionette shows on the
Paris promenades, or the operas of Rossini), this outward guise conceals a
deeper drama. The true drama of the *Théorie* is the drama of the observer
of moral nature who must constantly assure himself that his object is
graspable. A central passage of the *Théorie* expands on the epistemological
problem of observing human nature in the abstract:

> The observation of human phenomena, the art of perceiving the most
> hidden movements, the study of what little this privileged being may
> inadvertently allow us to divine of its consciousness: this demands
> both a certain measure of genius but also, paradoxically, its circum-
> scription. One must possess patience, like Muschenbrock and Spall-
> anzani before us, or Nobili, Magendie, Flourens, Dutrochet, and many
> others today. Then too, one must possess that expert eye [*coup d'œil*]
> that makes phenomena converge toward a center; that logic that ar-
> ranges them in orderly rays; that perspicacity that sees and deduces;
> that slowness that means we shall never discover one point on a circle

without observing the others as well; and that swiftness that can size up a thing from top to toe.

This multiple genius, possessed by a few heroic minds justly celebrated in the annals of the natural sciences, is much rarer in observers of moral nature. The author, whose charge it is to spread the enlightenment that shines from the high towers of science, must give his work a literary form, must make the most difficult doctrines interesting to read, must decorate the science. Thus he finds himself constantly ruled by form, by poetry, by the accessories of art. To be both a great writer and a great observer—Jean-Jacques [Rousseau] and the *Bureau des Longitudes* in one person—that is the problem, the insoluble problem.[119]

This reflection is squarely pitched at the standard repertoire of virtues of the good scientific observer. The insistence on heightened senses and a mind of near-infinite, almost godlike receptivity harks back to the imperative that natural science should attend to the world around it, something that had shaped the everyday practice of most scientists since the eighteenth century.[120] Above all, however, Balzac's critical diagnosis identifies a crisis state for the science of man. This crisis was evident in the estrangement between different observational genres, which were classified as either "scientific" or "literary," according to the descriptive styles they cultivated.[121] As is clearly demonstrated in the diverse French discussions of the physiology of movement, the problem was especially virulent in the study of gait.

But Balzac does not make the scientific observation of man's moral nature either primarily a psychological problem of attention or solely a problem of representation. Instead, the *Théorie de la démarche* simply complicates the dilemma: it directs the theoretician's gaze toward a phenomenon that is defined as ultimately *ungraspable*. Even the movement of inanimate bodies is called an "abyss" that threatened to swallow up reason: or as Balzac's fictional physicist Planchette declares in *La peau de chagrin*, "when it comes to movement itself, I say to you in all humility that we are powerless to define it. . . . Movement is an insoluble problem, like the vacuum, creation, or infinity; it confounds human thought, and the only thing man is permitted to understand is that he will never understand it."[122] In the *Théorie de la démarche*, Balzac picks up this theme in a parable of a madman and a geometer in which the madman hurls himself over the edge of the abyss while the geometer descends on a ladder with a ruler and other instruments, contenting himself with measuring the depth and temperature of the void. In declaring even the tiniest human movement to be such an abyss, Balzac signals the exceptional status of both his own theory and its author, who finds himself somewhere "between the measuring stick of the geometer and the vertigo of the madman." "This theory," he writes, "could only be the product of a man audacious enough to approach madness without fear and science without trepidation."[123]

The *Théorie de la démarche* further dramatizes the epistemological dilemma of the science of man—itself caught between regulated, scientific measurement taking and excursive, personal field reports—with the use of a very simple rhetorical device. Balzac scrupulously exploits the polysemy of the French word *démarche*, which may mean either a manner of walking or, by figurative extension, a methodical progress of thought or action.[124] The development of his "théorie de la démarche" can thus be conceptualized reflexively as a narrative that on the one hand functions as an illustration of a more general "theory of the march of ideas" (*démarche de nos idées*)[125] in its "step-by-step" observations and on the other hand depicts, in a series of episodes, the emotionally volatile relationship of the theoretician to his theory.[126]

The reflexive interlacing of outer and inner man described in the act of observation was consistent with the proposition in energetics that "a man could project outside of himself, through all of the acts that result from his movements, a measure of force which must produce some kind of effect within the sphere of his activity."[127] Following this "simple formula," movement was defined as a kind of primary principle that encompassed "thought, the purest action of human beings," as well as "the word, the translation of thought," and "walking and gesture," which were the "realization of the word with more or less passion."[128] With these ideas, largely derived from Franz Anton Mesmer's theory of animal magnetism, Balzac took up arms against the physiological mechanics of walking espoused in "specialized articles by physicians" who, "following Borelli's example," he found had "not investigated into causes so much as simply stated their effects."[129] Presumably Balzac alludes here to the nearly ritual return to the canonic *De motu animalium* that Roulin and Gerdy made in their articles for Magendie's journal. The very same return made by the theorist in the *Théorie de la démarche* is elaborately staged as a passionate dispute: "How lucky I was to find a Borelli on the quay! How light the quarto volume felt to me, tucked under my arm! With what fervor I opened it; with what haste I translated it! I cannot describe it. There was love in that study!"[130] In a quick reversal, however, his reading of Borelli's treatise turns to anger and disappointment: "Borelli indeed tells us why, if a man loses his balance, he falls; but he does not tell us why the man often does not fall, if he knows how to make use of a hidden force by transmitting an extraordinary *retracting* power to his feet."[131] The *Théorie* makes repeated appeals to the field of speculative energetics to expose the blind spots in vitalist physiologies of walking in relation to the existence of a "motive fluid" (*fluide moteur*), characterized by Balzac as an "unfathomable power of volition, the despair of great thinkers and physiologists."[132]

But neither could the theory of a "spiritual fluid" (*fluide animique*) underlying the "prodigious eloquence of the gait" ultimately provide secure footing.[133] The confrontation with various approaches to a physiology of walking that plays out in the *Théorie de la démarche* leads not to any progressive accumulation of results but rather to fresh contradictions that repeatedly circle back to the starting point. The observer comes finally to doubt whether the object of observation even exists: "I asked myself where movement begins. Well, it is just as difficult to determine where, in ourselves, it begins and ends as to tell where the sympathetic nervous system [*le grand sympathique*] begins and ends—that internal organ that has exhausted the patience of so many observers at present."[134] Balzac adopts the totalizing discourse of the science of man, which calls for the exhaustive observation of human beings in the act of movement on both the physical and moral planes, only to hand it its own defeat in the face of an omnipresent and banal but ultimately indomitable object. The theoretician of walking can only reach the inevitable conclusion: "*Nothing*: this will be the eternal motto of our scientific endeavors."[135]

Mechanicians of the Human Walking Apparatus: The Beginnings of an Experimental Physiology of the Gait

In the *Théorie de la démarche,* Balzac deplored in a provocative gesture the fact that scientists had spent more time studying the progress of the stars across the heavens than the progress of men across the earth: "Why has the walk of men drawn the shorter straw? Why has it been preferable to occupy oneself with the march of the stars?"[1] At the same time, in Germany, two young scientists became the first to use the instruments of astronomy to investigate human gait, when the physicist Wilhelm Weber trained a telescope from the astronomical observatory in Göttingen not on the skies but on the legs of his brother, the anatomist Eduard Friedrich Weber. In the monograph they published together, titled *Mechanik der menschlichen Gehwerkzeuge (Mechanics of the Human Walking Apparatus),* the Webers briefly described Wilhelm's observations of his brother's gait. "Using a telescope set at a distance," they wrote, "we observed and measured how high the leg was raised above the floor as the subject moved in its direction. The lift was close to one-ninth the length of the leg, and this differed little, whether the subject was walking fast or slowly."[2]

In its deliberate attempt to differentiate itself from every previous physiology of walking, the scene takes on a decidedly emblematic character. A familiar, everyday event is placed at a distance, no longer construed as a theoretical object seemingly resistant to rational observation solely through the use of chiefly textual practices or neologisms. The phrase *menschliche Gehwerkzeuge,* or "human walking apparatus," admittedly, was a coinage (and as such, works immediately to estrange us from an apparently familiar phenomenon),[3] but as researchers, the Webers pursued a new brand of experimentalism. In contrast to the French experimental physiologists, they did not pursue vivisection, and they set the question of the cerebral control of bodily movement aside completely. What they

sought to dissect was the complex act of locomotion itself: through a series of manipulations, they aimed to isolate the physiological processes
of walking and running in their pure forms as precisely calculable mathematical quantities.

Published in 1836, the *Mechanik der menschlichen Gehwerkzeuge*
quickly became an important reference work even outside of its narrow
field. In 1842, the *Conversations-Lexikon der Gegenwart* praised the book
as "a work that treats exhaustively a hitherto uninvestigated topic and
will be of equal interest to both anatomists and those who study the laws
of mechanics."[4] Around this time, Wilhelm Weber, working with Carl
Friedrich Gauß, also invented the electromagnetic telegraph; while that
event has gone down in the annals of history,[5] the Webers' gait study fell
into almost complete obscurity in the twentieth century.[6] Seen in the context of the early nineteenth-century attempts by French physiologists to
develop a vitalist mechanics of walking by revising and extending Borellian theory through clinical observation, the Webers' work appears as a
radical departure. What the Webers saw in the work of such scientists as
Barthez and Gerdy was not empirical research results but "on the contrary,
ideas dreamed up by this or that writer about [how] walking and running
might work, which he believes he can combine with incomplete observations that are based on nothing more than the way people *look* when they
walk or run."[7]

The new experimentalism of the Webers, which aspired to combine
a "physics of anatomy" with a "physics of physiology,"[8] predefined its
research object based on strict criteria. The wide range of cultural and
individual gait variations, gender-specific differences, and pathological
manifestations—all objects of primary interest to previous physiologists in
the nosographic tradition—were all immediately excluded by the Webers.
Whereas French physiologists had sought to articulate the semiotics and
mechanics of human walking within an anthropological framework—as
part of the new *Science de l'homme*—the two Germans identified "natural
gait" as an ideal research object for an investigation in pure physical mechanics. As we shall see, the Webers' ideal determination of gait followed
largely from their adoption of the objectives of industrial mechanics (*mécanique industrielle*) in which comparative measurements of human, animal, and machine locomotive performance were undertaken to expedite
transportation and organize military marching columns more efficiently
as well as to help correct existing anatomical illustrations. The first problem the Webers faced was thus one that had already been addressed in the
military literature of the late Enlightenment: how to accurately measure

the natural stride. The use of the pendulum in military science became a guiding principle for the Webers and their theory of gait. In the course of their research, they developed new procedures that married the spatial isolation and selection mechanisms of the laboratory with popular techniques employed to produce optical illusions, allowing the two brothers to reveal, as never before, the organized action of the human walking apparatus in isolation—quite literally—from the human itself.

NATURAL GAIT: AN IDEAL OBJECT FOR THE SCIENCE OF MACHINES

The Webers regarded gait as an ideal object for their physics of physiology. "In the human organism," they reasoned, "there are very few other movements so completely dependent on external forces that proceed so uniformly and are so little altered by the will or other life processes."[9] Their ideal determination of the natural gait as a self-regulated and uniform manner of locomotion represented a significant move away from clinical practice and similarly from an audience of medical specialists. In making this move, the Webers drew on the notion of the "natural gait" in which the foot travelers, naturalists, and soldiers of the late Enlightenment had located so many virtues. So much is clear from the following definition, which they relegate rather casually to a footnote: "By natural gait, we mean that gait that we adopt instinctively, without paying notice to each individual step, but simply with the intent of making forward progress. A man traveling for days on foot will always walk in the manner that is least fatiguing. His speed may vary greatly, but it usually remains constant if we do not consciously try to change it."[10]

The self-evident air of this definition may perhaps be attributed to the life experiences of the investigators: both were habitual walkers who (partly also owing to straitened finances) did not make even long journeys by coach.[11] They belonged, moreover, to a generation that came of age during the German campaign against Napoleon, when physical education for youth became a moral imperative, and the gymnastics movement headed by the "Turnvater" Friedrich Ludwig Jahn (1778–1852) was promoted to a paramilitary venture of national importance. This new patriotic movement occurred within the larger context of the reorganization of the German military, where the goal was to create a more mobile national fighting force on par with that of France.[12]

In 1816, Jahn and his assistant Ernst Eiselen, in their *Deutsche Turnkunst* (German gymnastics), had named four criteria that defined the art of good walking: "A good walker must combine both *speed* and *endurance* with *decorum*, and pay no heed to *terrain*: whether mountain or valley, sand or clay."[13] Since the Enlightenment, the practice of the foot journey in Germany had been associated primarily with an aesthetic and a moral sensibility. Jahn and Eiselen's definition recognized that foot travel was also increasingly a means of deploying a previously underexploited economic power. "*Endurance* in walking," Jahn and Eiselen wrote, "is achieved only through practice. Walking with children, daily when possible, gradually increasing the distance, and working up to long hikes and foot journeys, is the best way to practice. Great endurance in walking, including carrying the necessary pack, is a virtue whose worth is still underappreciated by many."[14] That the way the Webers defined natural gait conformed to the type of foot travel extolled by the gymnastics movement—by this time elevated to a collective practice—is apparent from the second edition of the *Deutsche Turnkunst*, which recommends the *Mechanik der menschlichen Gehwerkzeuge* as the sole reference for studying the "basic conditions for, and character of, the human gait."[15]

The Webers' selection of walking as their object of study thus had multiple roots in their historical moment: on the one hand, in a reassessment of foot travel as a cultural, aesthetic, and political practice, a trend that found multiple expressions in the Romantic music and literature of the German-speaking countries; and on the other, in a program of "technische" or "industrielle Mechanik"—as the French terms *mécanique industrielle* and *mécanique appliquée aux arts* were translated in Germany—that targeted industry, the military sciences, and the arts.[16] Although the Webers themselves never specifically referenced any such works, by addressing their work primarily to "amateurs" in the fine arts, "educated military men," and engineers, whom they challenged to invent new walking machines, they made their target audience unmistakable.[17] Their "clear depiction of the science of walking and running," accompanied by many illustrations, bears further witness to the exoteric orientation of their monograph, as do both the summary of their experimental results they include at the beginning and their decision to present their main tenets in language accessible to the layperson.

In defining the human walking apparatus as a mechanism that was unconscious yet regular in its operation (as quintessentially manifested in the long foot journey), the Webers became the first to fully marry Borellian iatromechanics (which they acknowledged extensively in a historical review) with the contemporary theory of machines (*Maschinenlehre*). The latter field was founded on efforts to measure the performance of humans, horses, and moving machines, dating back to the eighteenth century, above all in the fields of physics and military science and in athletic competitions.[18] Texts such as Franz Joseph Ritter von Gerstner's *Handbuch der Mechanik* (1831), informed by a principle of economy of movement where regularity, frictionlessness, and efficiency were the ideal, surveyed extraordinary physical performances by soldiers, runners, racehorses, and load-carrying workers and compared them to freight wagons, streetcars, and trains.[19]

Since Coulomb, performing friction calculations for the materials used in transport machines had been a crucial element in defining the forces of different moving bodies.[20] In the 1820s, however, when the first English steam locomotives began to compete with draft animals, the efficiency criteria used to evaluate means of transport appeared as the result not only of a movement toward increasing quantification and standardization but also of certain cultural ideals. What cultural historian Wolfgang Schivelbusch has called "the mechanization of motive power" assumed a steam locomotive on rails operating with completely uniform and frictionless motion: a new "mechanical ensemble"[21] that was contrasted, in idealized descriptions, with the allegedly deficient mechanics of animal locomotion:

> The animal advances, not with a continued progressive motion, but with a sort of irregular hobbling, which raises and sinks its body at every alternate motion of the limbs. . . . Even in walking or running one does not move regularly forward. The body is raised or depressed at every step of our progress; it is this incessant lifting of the mass which constitutes that drag on our motions which checks their speed, and confines it within such moderate limits. . . . With machinery, this inconvenience is not felt; the locomotive engine rolls regularly and progressively along the smooth tracks of the way, wholly unimpeded by the speed of its own motions.[22]

In this analysis, the arrangement of the bodily machine resulted in excess movement that inevitably limited speed and efficiency. Progress was also continuously impeded by frictions apparently arising inside it. This alone led early English steam locomotive apologists to see human and animal locomotion as unsuitable. Moreover, for various reasons, horses were considered to present significant hazards in traffic, something that had inspired criticism of stagecoaches since the late eighteenth century.[23]

The Weber brothers address the introduction of steam engines on public train lines in the opening pages of the *Mechanik der menschlichen Gehwerkzegue*. They invert, however, the rhetoric of technological progress according to which the human and animal locomotor apparatuses had been surpassed by the frictionless, regular movement of modern "mechanical ensembles":

> Regarding steam coaches, we know that they can only be used on a flat, hard, and horizontal or nearly horizontal route (such as a railway); but when going uphill the obstacles are so much increased that a steam wagon can overcome them only with difficulty, even on a moderate incline. On soft soil, the wheels dig in instead of rolling across the surface. Animals, on the other hand, and humans most especially, are never wholly impeded in their progress by any softness or unevenness of the ground, not even by a path that ascends or descends with considerable steepness. No other method of moving a load across the surface of the earth avoids the disadvantageous effects of friction and vibration better than walking or running; with no other method is the direction of motion so easily changed, with no other method can the tools so easily be adjusted to the various obstacles one seeks to overcome, as this one. We may thus confidently expect that once the mechanism of walking is truly understood, this knowledge will be of great benefit for the invention of new machines calculated for forward movement that will serve their designated purpose even in rough terrain where horse-drawn vehicles cannot be used and man must depend on the services of camels and other animals.[24]

Interestingly, the Webers' argument is actually aimed less at railbound steam locomotives than at a rival enterprise that is now largely forgotten: the so-called *Chaussee-Dampfwagen*, or steam-driven coaches, that were designed for use on normal roadways.[25] In 1834, the *Polytechnisches Journal* commented that "no other mechanical problem of recent years has captured the attention of the public as much as that of the *Chausseedampfwagen*. Its perfect solution would indeed not only have a wonderful effect on human traffic but would also very often be decisive in the grandest endeavors of our time: the construction of railways."[26] Steam coaches, which were built by numerous inventors (including the mechanician and optician Peter Wilhelm Friedrich von Voigtländer in Vienna; the Heaton brothers and Dr. William Church in Birmingham; and Sir Goldsworthy Gurney in London), stirred controversy in the 1830s both among

Figure 3.1 Dr. Church's steam coach. From the *National-Magazin der Gesellschaft für Verbreitung gemeinnütziger Kenntnisse*, no. 4, January 22, 1834.

expert observers and in the German daily papers (fig 3.1). Supporters declared the steam coach to be an unrivaled means of mechanized transport that would quickly "increase and speed up traffic without capital expenditure." Critics, meanwhile, argued for the superiority of railcars based on detailed friction comparisons of the two modes of transport.[27] Critics also emphasized the risk that steam kettles could explode on roadways owing to "the inevitable severe impacts to and shaking of all parts of the machinery," and they warned against their use in passenger transport.[28] Various public attempts to match the speed of railway locomotives using the new steam coaches drew crowds of onlookers but also resulted in a succession of spectacular accidents. After 1835, "railless steam coaches" were regarded—at least on the Continent—as an "adventuresome product of idle genius," unusable in real traffic.[29]

The Weber brothers joined the numerous critics of steam coaches, but in contrast to those who championed steam trains instead, the Webers believed that the most perfect, most functional locomotive machine was the organic locomotor apparatus of animals—especially that of humans. They advocated its use as a model for new steam-driven machines that would walk "on two, four, six or more legs."[30] They envisioned marrying the special capabilities unique to legs—also lauded by Jahn and other walking enthusiasts—with the new, motorized power of the steam engine to solve the problems faced by a road-bound steam coach on steep or uneven terrain.[31] "A knee-shaped rod," they wrote, "could be attached to the steam coach, mounted at the top front, pointing downward and backward, and extended by a steam piston located in the knee joint; it would brace itself against the ground to push the coach forward, more or less as a human would push [themselves forward]."[32]

The Webers' utopian vision of a walking steam engine may seem bizarre today, but within the fragile constellation of cultural and technological conditions that prevailed during this transitional era, it appears as a perfectly plausible project even if it was never realized. If it remained purely a work of the imagination, this was surely also because, on both a scientific and a cultural plane, it wedged the ideal mechanism of the human gait somewhere between two paradigms in the theory of machines: on the one hand, the traditional machines of classical mechanics, used to revive a physics of living bodies that went back to Borelli; on the other, the new machines of thermodynamics, which figured as objects of risky experiments not only in science but also in public transportation.

THE WEBERS' PENDULUM

In contrast to mechanical engineering, which was most nearly a utopian horizon for the Webers in terms of applying their theories, military gait research offered a much closer connection to an existing practice of measurement taking and experimentation. From the theory of machines, the Webers took their definition of the human body as a kind of coach consisting of a load (torso, head, and arms) and moving supports (the legs). This

enabled their crucial selection of only "those parts of the skeletal frame
that are the foundation of the human walking apparatus" as their experimental object.[33] The isolation of this heretofore unobserved object was
represented visually in a schematic drawing that depicted in silhouette
only those skeletal elements deemed necessary for locomotion, omitting
such extraneous parts as the vertebral arch, ribs, shoulder blades, and so
forth (fig 3.8).

Their aim to isolate the mechanism of gait and study it independently
of external factors led the Webers to combine experiments on the physical
properties of the "walking apparatus," which they performed on cadavers
in an anatomical cabinet, with observations of walking *in actu*, taken in
a simple laboratory environment shielded from the elements.[34] The laboratory consisted of a straight track laid out in an "attic space with a long
expanse of flat floor. . . . Large spaces were left at both ends of the track,
partly to enable the omission of the initial steps from the measurements,
partly to avoid the inevitable disruptive influence of a looming wall on the
final steps of the person approaching it."[35] Although some experiments involving faster modes of locomotion (such as the *Sprunglauf* [leaping run])
had to be conducted outdoors, most of their gait-measuring experiments
took place in this indoor space:

> Measurements were taken as follows: the walker, holding a [stop]watch,
> took up a position six to twelve paces away from the start of the track;
> walking forward, he started the stopwatch, which had been held still by
> a catch, at precisely the moment he entered the track and stopped it just
> as promptly at the moment his foot reached the end of the track. . . . At
> the same time, he counted the paces taken during this time span. The
> various actions required to take the necessary measurements could
> certainly have been divided among several persons, but here they were
> performed by the walker alone, because only he is in the position to
> perceive them with perfect simultaneity. As experience has taught us,
> when a second person takes the watch, a small amount of time elapses
> between the signal given by the first person and the starting of the
> watch by the second, and this time may be of varying lengths, render-
> ing the observations unreliable. Combining several actions in one per-
> son, given a little practice, does not affect the experiments.[36]

As this experimental setup—with all necessary measurements assigned
to a single person—makes clear, the Webers' work picked up the thread
of the military gait research begun in the late eighteenth century. Their

method of measuring pace lengths along a measured track, which they applied in various experiments, was certainly derived from practices in use since Guibert's *Tactique*. They complicated it with an epistemological problematization of the precision of multiple individual observers, a topic of concern for a number of scientific disciplines in the early nineteenth century, but primarily astronomy.[37] To be sure, nowhere do the Weber brothers make specific reference to the literature of military science, but they are known to have had contacts who were in a position to give them access to military research. One such contact was the Berlin astronomer Johann Franz Encke, who in 1813 interrupted his studies with Gauß in order to participate in the German campaign and in 1834 sent Wilhelm Weber various observations that had originated with "educated military men" and were relevant to the study of gait.[38] It is only to be expected that an "educated military" would exhibit interest in research that was based on precise measurement taking and that postulated that gait was governed by natural laws based on engineering precepts, in particular the principles of external resistance and friction, given that this topic had been discussed within military science at least since the publication of Carl von Clausewitz's famous treatise *Vom Kriege* in 1832. In von Clausewitz's analysis, a "tremendous friction, which cannot, as in mechanics, be reduced to a few points,"[39] was the major culprit preventing the application of theoretical military science to the vicissitudes of actual battle.[40] Friction, a concept borrowed from physical mechanics, was pitted against the popular but oversimplified image of the army as a great "military machine": "we should bear in mind that none of its components is of one piece: each part is composed of individuals everyone one of whom retains his potential of friction."[41] Tellingly, von Clausewitz illustrates his general concept of friction by describing the reality of early nineteenth-century travel in horse-drawn carriages, an experience always fraught with unexpected obstacles and difficulties:

> Imagine a traveler who late in the day decides to cover two more stages before nightfall. Only four or five hours more, on a paved road highway with relays of horses: it should be an easy trip. But at the next station he finds no fresh horses, or only poor ones; the country grows hilly, the road bad, night falls, and finally after many difficulties he is only too glad to reach a resting place with any kind of primitive accommodation. It is much the same in war. Countless minor incidents—the kind you can never really foresee—combine to lower the general level of performance, so that one always falls far short of the intended goal.

Iron willpower can overcome this friction: it pulverizes every obstacle, but of course it wears down the machine as well.[42]

In contrast to this point of view, which responded to the principal unpredictability of all organic and mechanical movements by making the "tact" and experience of officers and generals the ultimate authority,[43] the Webers wanted to demonstrate that the laws of human movement could be precisely calculated—but not only through the use of a variety of measuring and monitoring tools. They also sought to show that the oscillations of the pendulum, long used in the military to set drill paces, were, in the natural gait mechanism, identical with the motion of the legs. As discussed in chapter 1, military scientists of the late Enlightenment used seconds pendulums to coordinate the marching of soldiers with a prescribed beat—not so much for reasons of financial economy but rather because they believed a mimetic effect would lead recruits to match the motion of their legs to the motion of the pendulum.[44] By drawing an analogy between the body's *Gehwerkzeuge*, its "walking tools" or "apparatus," and the pendulum, the Webers created two memorable images: the image of a powerful gravitational force that seemed to propel the body forward of its own accord, and the image of an ideal mobile machine subject to neither internal nor external friction.[45] To prove their theory that the leg acted as a freely swinging pendulum, they measured the periods of oscillation of their own legs and those of other test subjects; and they took comparative measurements of the legs of human cadavers that they swung from ropes (table 18 in Weber, *Mechanik*; reproduced in fig. 3.2).

In an experiment on cadavers performed at a congress of German naturalists in Bonn in 1836 in the presence of multiple witnesses, Eduard Weber demonstrated that only the ambient air pressure, not muscles and tendons, held the human leg close to the trunk. Weber arranged the torso of a cadaver so that one leg could swing freely, and then drilled a hole into the side of the hip socket, demonstrating how the leg dropped when air entered the joint.[46] This experiment (which also features prominently in the *Mechanik der menschlichen Gehwerkzeuge*) made it possible to see the human leg as the equal of the best precision instruments of physics, since Nature's own design for the hip joint already eliminated all possible sources of friction:

> The leg is so very mobile because it is carried by the air, and its great
> mobility, as it hangs from and is borne forward by the trunk, allows it
> to swing like a pendulum, and starting from the inclined position at

Tabelle 18.

Vergleichung der Schwingungsdauer des Beins am Leichname und an lebenden Menschen.

No.	Länge des Beins	halbe Dauer einer Schwingung	Bezeichnung des Beins
	m		
1.	0,831	0″,370	ein exarticulirtes frei auf-gehangenes Bein
2.	0,866	0, 371	desgleichen
3.	0,831	0, 366	ein bis auf die Kapsel vom Rumpfe getrenntes Bein
4.	0,831	0, 355	ein unverletztes Bein am Leichname
5.	0,860	0, 346	ein lebendes Bein an schlaf-fen Muskeln frei herab-hängend
6.	0,860	0, 332	ein lebendes Bein beim Gehen auf der Ferse
7.	0,860	0, 323	ein lebendes Bein beim Gehen auf dem Ballen

Figure 3.2 Table 18 in Wilhelm and Eduard Weber, *Mechanik der menschlichen Gehwerkzeuge* (1836).

which it was lifted from the ground, to arrive by no other force than its own gravity at a vertical position, and from there to move almost as far to the opposite side (toward the front) as it had traveled from the first side. This organization of the hip joint is reminiscent of many physical instruments and ingenious methods that have been devised and applied to produce perfectly free rotation in heavy bodies and which consist in offsetting the weight of the body and preventing the friction of solid bodies against one other (which is the true enemy of rotation); in short, as one usually says when speaking of these instruments: *the weight of the body is equilibrated*. To the same end, the *weight of the leg is equilibrated*.[47]

Perhaps surprisingly, the Webers' theory, which conceptualized the mechanics of walking as a physical process carried out by frictionless precision instruments, initially met with less enthusiasm from the military

than from other natural scientists. Alexander von Humboldt, for instance
(a personal friend of the Webers who praised Eduard, particularly, as the
"more ingenious" of the two and "almost as knowledgeable in mathematical physics as in physiology"[48]), discovered in the *Mechanik der menschlichen Gehwerkzeuge* the explanation for a phenomenon he had repeatedly experienced on hikes in the Andes in 1802. Lecturing to a naturalist
congress in Jena in 1836 about his failed attempt to scale Mount Chimborazo, Humboldt speculated that the "strange fatigue one experiences when
walking at very high elevations" was a consequence of the thinner air at
high altitudes.[49] He asked the Weber brothers to perform another experiment in his presence to corroborate their theory. The men all convened
in the physical laboratory of the Berlin professor Gustav Magnus along
with the physiologist Johannes Müller. For their experiment, they bisected
a "fresh pelvis with femurs" through the sacrum, and trimmed it such
"that one could suspend the hip joints conveniently under the bell of an
air pump."[50] They drilled two holes into the hip joint: one above the joint
to suspend it from a rope, and one below, to which they attached a two-
pound weight to stand in for the weight of the missing lower leg. They
then severed the membrane encapsulating the joint and lowered the air
pressure in the bell, causing the head of the femur and the suspended
weight to drop. Letting air into the bell, conversely, caused the ersatz leg
to rise—thereby, in the eyes of the scientists, delivering not only proof
of the Webers' theory but also some useful practical knowledge for naturalists hiking in the mountains. Lower barometric pressure at high altitudes would produce higher tension in muscles that normally remained
relaxed throughout their swinging, pendulum-like motion; this explained
the "discomfort and inconvenience for walkers" and the great fatigue that
Humboldt had described.[51]

THE APPEARANCE OF TRUTH IN REPRESENTATIONS
OF GAIT: THE WALKING APPARATUS AND
OPTICAL-ILLUSION MACHINES

With their new theory of walking and running, the Webers wanted not
only to solve practical problems in military science and explain the physics behind the high-altitude fatigue experienced by mountaineers but also
to subject depictions of locomotion in the fine arts to scientific correction.
The discussion of anatomical accuracy in the arts was by no means new,[52]
but in the wake of research performed in the nineteenth century on moving bodies, it took on a new character. Earlier, in his lessons on art anat-

omy, Gerdy had insisted that artists should have scientific instruction in the correct observation of moving bodies. For him, this consisted largely in the compilation of an extensive catalog of errors in which numerous masterpieces were submitted to merciless scrutiny. In particular, Gerdy took issue with the widespread view that held up classical Greek sculpture as evidence of exact anatomical observations. He presented himself as the better observer of human bodies thanks to his method of subdividing the muscles and joints into multiple numbered elements (fig. 3.3).[53]

The Webers picked up the threads of this critique, but using their innovative combination of anatomical and physiological experiments, they took it in a new direction. In the area of anatomy, Eduard Weber experimented with the limbs of cadavers, with tendons either intact or ruptured, to see how far they could bend, stretch, and twist in different directions. To create a visual record of the curvature of the articular surfaces of the joints, the vertebrae, and the incline of the pelvis, he used a quite singular imprint technique, sawing through still-fresh bones in the direction of motion and having a printer stamp images on paper with them, "in the manner of a woodcut block." "By making plaster molds of the bones which we then used to make stereotypes," he wrote, "we could reproduce these imprints at our convenience and even include them in this publication. They are the most true-to-nature images available" (fig. 3.4).[54]

The Webers believed these "most true-to-nature images" of the human anatomy—Eduard called them the "imprint of the object itself"[55]—were superior to even the "best depictions of human skeletons." The very first plate in their study discloses errors in an older illustration of a male skeleton used by the Leyden anatomist Albinus (cf. figs. 3.5, 3.6).[56] Their decision to place this most representative image at the head of their work was deliberate. Like most images of its kind, the anatomical drawing used by Albinus represented a compromise between anatomist and artist.[57] Seeking to ensure that the skeleton would be rendered correctly on paper, Albinus had required his draftsman, Jan Wandelaar, to use a complicated setup of Albinus's own devising. To avoid perspective distortion, Wandelaar was positioned forty feet from the skeleton. The skeleton itself stood behind a wooden frame subdivided into squares, and the same pattern of squares was traced onto Wandelaar's drawing paper. This guaranteed that the skeleton would be reproduced in its correct proportions. Wandelaar was given a free hand in filling in the background decoration, for which he used a number of conventional emblems (see fig. 3.5, in which the walking skeleton is placed before a monument set in a landscape).[58] Significantly, the Weber brothers have no comment on Albinus's rather

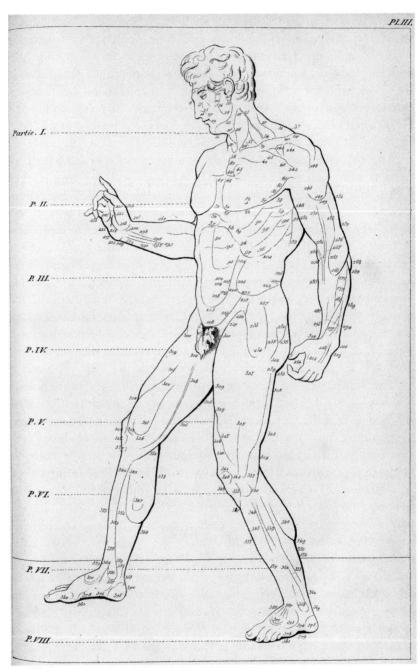

Figure 3.3 Plate 3 in Pierre-Nicolas Gerdy, *Anatomie des formes extérieures du corps humain* (1829).

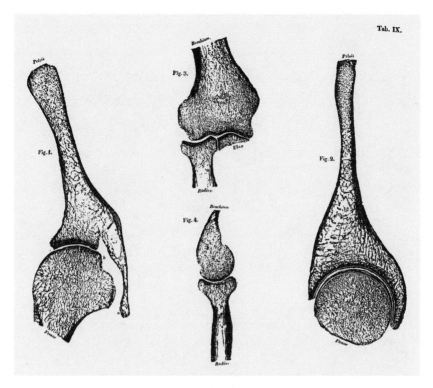

Figure 3.4 Plate 9 in Wilhelm and Eduard Weber, *Mechanik der menschlichen Gehwerkzeuge* (1836).

complicated arrangements. They reserve their criticism solely for Wandelaar, who had incorrectly represented the inclination of the pelvis and the curvature of spine as compared to observations of living persons. In their corrected illustration, the Webers also remove the landscape around the skeleton (fig. 3.6).

Proceeding on from this critique, the Webers argue that artists seeking to portray human movement should use the technical images produced by experimental science as their primary resource:

If the science of running and walking can correctly predict the positions of the trunk and the individual limbs that arise simultaneously in the course of such locomotion, then a corresponding drawing will impress even a layperson with the faithfulness of its depiction: for it is easier to recognize the correct position of the torso and the various limbs, as required for whatever purpose, than to discover it for

TAB. III.

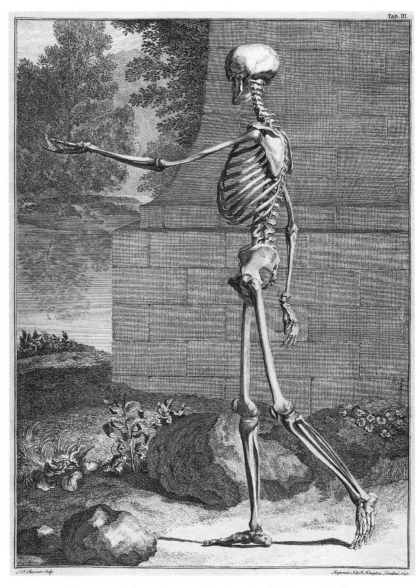

Figure 3.5 Skeleton, from Bernhard Siegfried Albinus, *Tabulae sceleti et musculorum corporis humani* (1747).

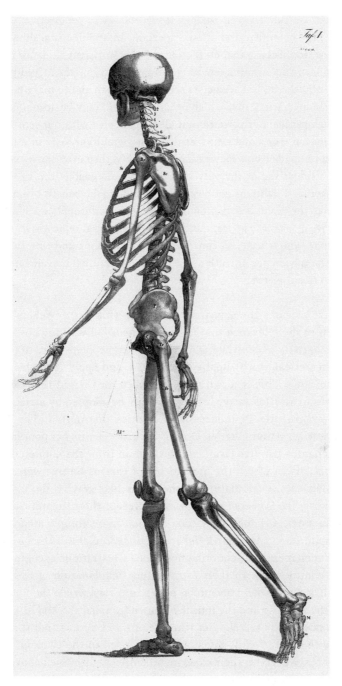

Figure 3.6 Corrected version of Albinus's skeleton. Plate 1 in
Weber, *Mechanik der menschlichen Gehwerkzeuge* (1836).

oneself. We can draw an analogy to the science of perspective, which is so very helpful to the artist, especially in architectural drawings. No one would claim that correct intuition is sufficient guidance for an artist, . . . And even if a few artists of rare ability might indeed hit on the truth, the task of science is precisely to teach and to make achievable through study that which otherwise could only be accomplished through genius. But if we cannot clearly perceive with our senses the positions of trunk and limbs, as they arise simultaneously in walking and running, because they pass by so quickly, then even the most able artist is deprived of the very thing that would enable him to recognize actual conditions as they unfold, without the benefit of science, through the direct observation of nature. Just as the application of perspective creates the appearance of truth in pictorial representations of distances and objects, so can the theory of walking and running impart the appearance of truth to the human and animal movements portrayed in our pictures.[59]

To give pictures the appearance of truth: this is the Webers' clearest statement of the relevance that their new theory of walking possessed for the arts. Artists, whom they addressed as "laypeople," were urged to accept a theoretical truth supplied by science and apply it to their figural representations. This recognition of the need for revised images of moving figures in an illustrative form that could be adopted by artists already figured prominently in industrial mechanics instruction. The Parisian mathematician Baron Charles Dupin, for example, in his popular industrial mechanics lectures (available in German from 1826), offered a series of illustrations in which the movements of various figures were adjusted in accordance with calculations of their centers of gravity (fig. 3.7).[60]

Likewise, the Webers found a way to present their theoretical laws in a vivid manner, although they took a much more complicated approach than Dupin. In the *Mechanik der menschlichen Gehwerkzeuge*, they include a supplement that contains numerous illustrations accompanied by detailed commentary. In their introductory "explanation of the illustrations," they divide their seventeen plates into two series: the "larger pictures of the skeleton and the muscles, drawn from nature, and direct prints of the bones," and "the smaller illustrations and figures that function to elucidate the text."[61] This division corresponded to the twofold approach taken by the Webers in their experiments already discussed above: on the one hand, they were pursuing a physics of anatomy that investigated the "bodily machine designed for walking" when at rest (i.e., in cadavers); on

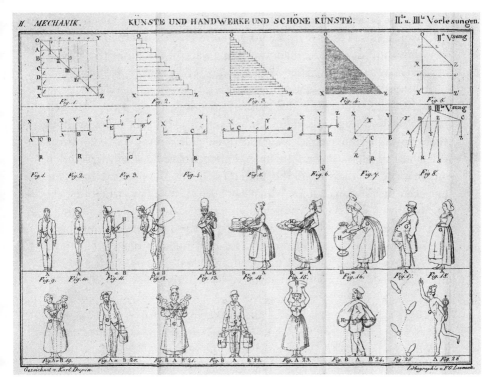

Figure 3.7 Plate from Carl Dupin, *Geometrie und Mechanik der Künste* (1826).

the other, a physics of physiology whose object was the walking machine in motion (in living subjects).[62]

Each series of illustrations in itself, however, is also highly heterogeneous. In the first series, anatomical drawings of skeletal elements are juxtaposed with the "bone prints" described above. The drawings here are primarily intended to extend and complete the prints, illustrating otherwise invisible elements such as the tendons of the knee. The second series of illustrations, which the Webers state are meant to clarify the main text, is even more disparate. It mostly consists of pictures of human figures that illustrate the Weberian theory of walking while also demonstrating how to draw such figures. Most are meant to give a "visual illustration" of some of the experiments performed by the authors. The striking increase in diversity within this series, however, reflects a shift of perspective. In the first series of images, objects inscribe themselves into the book, with each imprint becoming a kind of stand-in for a "real specimen" in an anatomical investigation.[63] The second series, by contrast, depicts the authors' experimental practice and some of the theoretical axioms derived from it, and here the experimentalists and their instruments step into the pictures themselves—for as we have already seen, their study of living walking machines (the "physics of physiology" approach) relied almost exclusively on self-experimentation.

In this act of representation, therefore, a problem of reflexivity inevitably arose. How were the brothers to bridge the gap between self-impressions of dead walking machines and depictions of living walking subjects, whose movements were mostly recorded in the form of numerical tables of measurements? The Webers contrived a striking solution, one that could only work in the medium of pictorial representation. They mapped a succession of steps (which had been executed by one brother acting as a test subject) not onto a human figure but onto either a single leg or a skeleton (which were shown walking at various speeds), reproducing the individual stages of movement as a continuous sequence of images (figs. 3.8, 3.9). To prevent this unfamiliar mode of illustration from being misunderstood, they comment at the very beginning of their text:

> Those parts of the bony frame that make up the basis of the human walking machine we will show in silhouette; but we will leave out other parts such as the vertebral arch, the ribs, shoulder blades, etc., that have no direct effect on the walking mechanism. Furthermore, the spinal column and the ends of many bones are depicted in lengthwise section, because this gives us a much more correct notion of the curvature of the articular surfaces and the positions of the joints and their pivot points.[64]

Their walking skeletons, in other words, are reductions—extracts, as it were, from the complete form—that clearly present the "naked" theoretical truth en route to a purely mathematical definition. The Webers made the leap from anatomy to theory by simply vaulting over any visual depiction of their physiological research. To be sure, the text fills the gap in a rather amazing fashion. In an introductory "clear presentation," the brothers summarize the main points of their theory of walking as follows: "Everyone will instantly perceive that the person in figure 12, shown at 1:20 scale, is walking much faster than the person in figure 13. In fact, figure 12 is a sketch of a person with a stride length of 700 millimeters, while figure 13 is a sketch of a person whose stride length is only 600 millimeters"[65] (fig. 3.9). Both pictures ostensibly show the same person walking. That person, however, appears only as a "sketch"—by which the Webers mean not a "rough draft" but rather a partial view, one that shows only

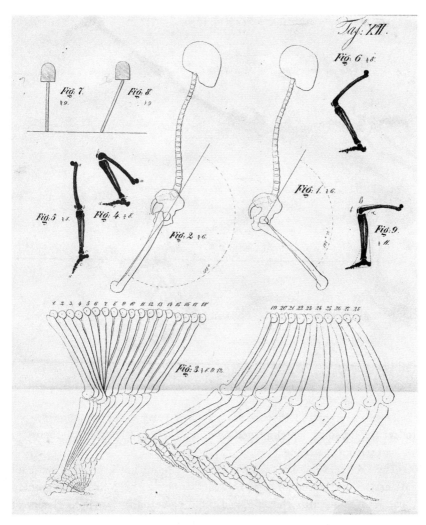

Figure 3.8 Plate 12 in Wilhelm and Eduard Weber, *Mechanik der menschlichen Gehwerkzeuge* (1836).

what is relevant for their theory. What we see, therefore, is not a drawn rendering of an empirical observation—as the explanation just quoted might suggest—but a representation of the unconscious mechanism of locomotion.[66] The locomoting skeletons thus build a visual bridge between the Webers' anatomical cadaver experiments and their physiological self-experiments. The drawings pull the unobservable mechanism of walking, wherein gravity moves the feet like pendulums, into the picture, and they

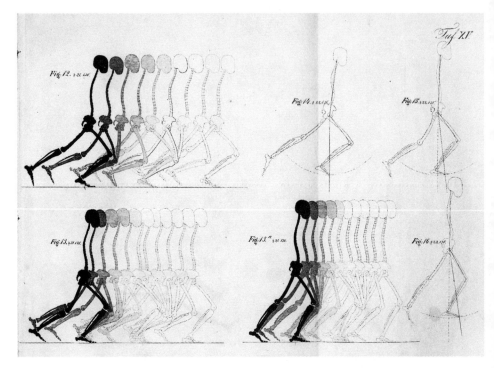

Figure 3.9 Plate 15 in Wilhelm and Eduard Weber, *Mechanik der menschlichen Gehwerkzeuge* (1836).

redact the muscles, which played no crucial role in the brothers' theory of walking, from the scene.[67]

The skeletal sequences were meant to do more, however, than just educate artists on the theory of walking. Ultimately, the Webers also granted these images a status that allowed them to actually test their theory of locomotion. To do so, they employed a "familiar artifice": an array of pictures of a moving skeleton drawn on a strip of paper in a rotating projection device, which could "produce the effect of a walking figure, entirely naturally."[68] The apparatus was based on the same principle as the stroboscopic disks, or "magical optical disks," invented in 1833 by the Austrian mathematician Simon Stampfer and marketed successfully by his contemporary, the Belgian inventor Joseph Plateau, as "phenakistiscopes" (fig. 3.10).[69] This extremely simple device exploited the afterimage effect described by Faraday in which pictures viewed in quick succession merge in the brain owing to the eye's inability to resolve the individual images quickly enough (fig. 3.11). On a disk with perforated slits, sequential

Figure 3.10 Stroboscopic disks, from Simon Stampfer, *Die stroboskopischen Scheiben oder optischen Zauberscheiben* (1833).

Figure 3.11 Phenakistiscope by Joseph Plateau (1833).

drawings were made of "any visible object, whether a machine in motion or humans or animals executing any kind of action or movement." Viewers watched the drawings in a mirror as the disk rotated, producing the illusion of an "animated picture":[70]

> Nearly all movements made by machines, like the actions of walking and running in humans and animals, and many other human actions and activities besides, occur over a period of no more than 2 or 3 seconds and thus are well suited to our method of representation. The object in motion has traveled through one period at the precise moment of its return to a previous position: i.e., a rotating wheel resumes the same position again with each spoke, hence one may include one or more spokes in a single period. For a person walking, each step is one period, if one ignores the difference between left and right foot, etc.[71]

As early as 1834, numerous commercial variants of this popular optical-illusion machine were also to be found in England and France. One such variant, invented by William Horner, was called the daedaleum (later also known as a zoetrope or *Schnellseher*). Inspired by an idea of Stampfer's, the daedaleum arrayed a series of pictures around the inside of a cylindrical viewing device (fig. 3.12).[72] The Webers exploited this improved version of a "magical disk." As they wrote, "We worked out entire systems of these figures according to theoretical prescriptions, and using a cylinder that turned at a measured speed, we presented them to the eye in quick succession, so that together they produced a remarkable effect of illusion: as if one really saw a figure walking or running, and doing so, indeed, in the most natural manner."[73] The advantage of the new device over the earlier rotating disks is obvious. The phenakistiscope could accommodate only one viewer at a time, who regulated the speed of the disk in front of the mirror. With the cylinder, however, the moving figures could be demonstrated for a larger audience. The observers also assumed a different position vis-à-vis their object: they no longer had to bring their eyes to the disk but could place themselves further away from the apparatus, while the cylinder itself (assuming it was operating at the proper speed) produced the desired effect of natural movement.[74]

Thus, the theoretical truth that the Webers demonstrated in the *Mechanik der menschlichen Gehwerkzeuge*, and that was intended to help artists improve their pictures of figures in motion, was communicated to audiences with the help of a popular optical toy. But the relationship also moved in reverse: the principle of the stroboscopic device itself

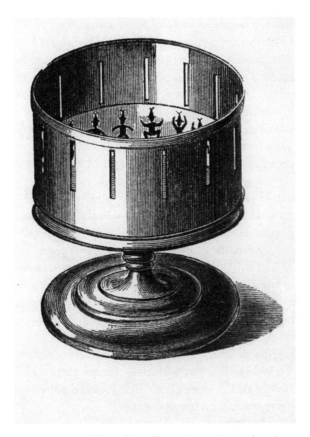

Figure 3.12 Daedaleum by William George Horner (1834).

helped establish the new object of the "walking apparatus" (which could refer to any real or fictional locomotion machine) on a theoretical level: "Had one never seen a human walk or run and only knew the ratio of his limbs, with the aid of theory one could still reach an understanding of these movements that would be quite consistent with experience and be able to predict any part of the process."[75]

INDIVIDUAL VARIATION IN UPRIGHT GAIT

The *Mechanik der menschlichen Gehwerkzeuge* achieved a foothold in German physiology thanks largely to the fact that its conclusions were reprinted in the second edition of Johann Müller's *Handbuch der Physiologie* (1840). "The discoveries of these scientists," Müller writes, "have finally brought a rational acuity to the physics of locomotion."[76] Müller's assessment proved of decisive importance for one of his students, the Frankfurt anatomist Hermann von Meyer (1815–1892), who devoted himself to further experimental investigation of the walking apparatus. In 1844, Meyer took up a professorship in Zurich, where his major focus became the mechanics of the human skeleton, in particular the anatomy of the foot (helping to initiate far-reaching reforms in shoe construction in the process).[77]

Meyer presented the results of his studies of the "locomotor apparatus" in a series of articles that appeared in Müller's *Archiv für Anatomie, Physiologie und wissenschaftliche Medicin* in 1853. In the articles, Meyer undertakes a critical revision of the Webers' *Mechanik*. In his view, the "most natural and unforced" posture was the "erect military posture, with perfect symmetry on both sides of the body." "All other stances," he writes, "are only modifications of this simplest form and arise by adding some new element to it: namely, the exertion of specific muscle groups."[78] In a series of tests in which he dropped a plumb line beside a person standing upright, Meyer corrects the point at which the Webers had located the center of gravity of a stable, upright torso. According to Meyer, the Webers' data leads to a "forced and unstable posture that one cannot bear to hold for very long."[79] One of the main reasons for the discrepancy, he claims, was that the Webers had not performed identical measurements across multiple individuals but instead calculated an average value from several measurements of *"one* well-formed person."[80]

The Webers' preference for using anatomical cadavers of supposedly ideal proportions was, in fact, congruent with their research strategy, as they had striven to exclude individual and cultural influences from their study. Meyer, by contrast, thought it was precisely the problem of "individual variation in upright gait" that was the true challenge for anatomy. He noted: "it is obviously completely impossible to identify one specific manner of walking as *the* walk, Κατ ἐξοχήν [kat' exochēn]. And this is clearly only natural, considering that 'gait' is no given thing but in general only consists of a forward movement of the body aided by the legs—that such a complicated apparatus as the leg can be used in the most varied ways to achieve this goal—and that every person 'walks' just he can and as it pleases him."[81]

This position represented a fundamental shift of perspective. No longer could natural gait be an ideal object for the field of mechanics, the product of a frictionless walking apparatus. Now it appeared more likely to be an impossibility, or at least an extreme improbability, in a diverse world of gaits both healthy and pathological, for which the legs functioned as "aids." In this manner, Meyer transposed the knowledge of the numerous variations in gait from clinical semiotics (above all as it had been articulated in French physiology beginning in the early nineteenth century[82]) into experimental anatomy.

On a practical level, he accomplished the transfer by combining anatomical measurements of the body, taken with plumb lines and compasses, with a procedure already in use by forensic scientists (both real and fictional) to describe individual physiognomies of gait: namely, footprint analysis. By combining these approaches, Meyer could extend and ultimately revise the prevailing theory of the walking mechanism as a pendulum.[83] He started with what he called the "walking line," which could reveal individual peculiarities of gait: "In any manner of walking, the footprints fall in a way that permits a straight line to be drawn that will pass through every heel print. I will call this line the 'walking line.' Whether this line runs closer to the outside of the heel prints, closer to the inside, or through their centers depends on the individual character of the gait."[84]

Meyer's attempts to describe the "individuality of the gait" using walking lines employed a quite simple procedure: footprints were "obtained by walking with wet, bare feet on the floor of a room."[85] Hardly by coincidence, it was also around this time that a new technique emerged in forensic science that used footprints left at crime scenes as a means of criminal identification. Beginning around 1850, forensic physicians writing in the *Annales d'hygiène publique et de médecine légale* published a number of methods for preserving and measuring footprints.

These methods followed one of two approaches. The first was to transform the footprint into a tangible object that could be produced as evidence in court. To this end, a number of techniques were developed to reproduce transient footprints in a solid, three-dimensional form that could serve as a surrogate for the originals.[86] The second approach led to the development of another method that closely corresponded to the clinical analysis of footprints and may be traced further back in time. In those cases where a murderer had tracked barefoot through a victim's blood, it was argued that the footprints could serve as a "signature," a kind of writing that was unambiguously individual and permitted the murderer's positive identification.[87] In this special case, suspects were required to walk through blood to leave footprints that could be compared with the originals. Although the procedure did aim to identify specific individuals, the method relied on a general typology of various foot shapes and forms (male or female, adult or child, flat, deformed by shoes, etc.).[88]

The revisions Meyer made to the Webers' theories after 1850 based on his anatomical experiments as well as the budding discussions in forensic science about the preservation of footprints underscore the great extent to which the "walking line," composed of a series of consecutive footprints, was viewed in the human sciences as an expression of the individual. The footprint, which initially had mostly figured in an essentially divinatory physiognomy before it featured in detective fiction as the clue par excellence,[89] emerged in the latter half of the nineteenth century as an integral element in a number of innovative methods for the empirical reconstruction of diverse gaits.

The Rise of Graphical and Photographic Methods: Locomotion Studies and the Predicament of Representation

In an 1855 paper prepared for the Académie des sciences in Paris, the French neurologist Guillaume-Benjamin Duchenne presented a number of observations he had made of men and children who exhibited various pathologies of gait. In the case of a cart driver whose irregular body movements Duchenne first observed from afar in the streets of Paris, closer interrogation revealed that a work accident had led to a paralysis of the flexor muscles of the thigh. Duchenne described how laboriously the man now walked: "he lifted and threw the limb that should be moved forward with great effort at every step by raising the corresponding shoulder and by imparting a rotating motion to the pelvis, from back to front, over the opposing limb."[1] On the basis of this and other cases of muscular atrophy, Duchenne argued that the essential condition for normal walking was not the action of gravity on the legs, as the physiological theory of the Webers claimed, but rather the contraction of the flexor muscles. These few clinical observations were, in his view, sufficient to refute the Webers' mechanical analysis of locomotion and their theory of the swinging leg as a pendulum. Duchenne concluded that the simultaneous contraction of the thigh, leg, and foot flexors was "the real productive cause of the movements of the lower extremity" in this second phase of walking and that "the action of gravity plays a very unimportant role in the physiologic oscillation of the lower extremity."[2]

Even before Duchenne's critique, the Webers' experimental work on the mechanics of the human walking apparatus had received a mixed reception in France. In the 1850s, however, it came under heavy attack.[3] Inspired by Duchenne's observations, the physician and prolific writer Felix Giraud-Teulon, who had trained at the École polytechnique in Paris and was thus well versed in mathematics, undertook a systematic refutation of the pendulum theory. Giraud-Teulon concluded that the Webers' investigation was flawed from the start by a number of preconceived notions, most notably the idea that a person was a "locomotive made of fixed elements" (*locomotive à éléments fixes*),[4] a "peculiar, speculative idea"[5] that had led them to make ill-conceived experiments and faulty calculations.

Over the following decades, French research on locomotion largely oscillated between the approaches taken by these two critics. The first approach, physiological and experimental in nature, emerged, despite its critique of the Webers, as a more sophisticated version of their mechanical view of the "human walking apparatus." Its most famous representative was the French physiologist Étienne-Jules Marey, who began studying locomotion shortly after his election into the Collège de France in the 1870s. His experimental project was not only steeped in the anatomical tradition that had begun with Borelli but also, like its German antecedents, strongly connected to the fields of industrial mechanics, the military, and the fine arts. In contrast to the Webers, however, Marey wanted to experiment only on living bodies in motion and to create a new open-air experimental setting that would allow for comparative studies of human and animal locomotion.[6]

The second approach, in the wake of Duchenne's work and the gradual establishment of neurology as a discipline, led to the progressive development of a clinical semiotics of human walking. As the following chapter will discuss, the development of Marey's new *physiologie graphique*, "graphical physiology," in many ways stood in contrast to the use of traces to diagnose gait pathologies in clinical medicine. In order to understand developments in the last quarter of the nineteenth century, we must turn first to the field of hippology, where studies of equine gait sparked a number of controversies in French society, military science, and the arts. The invention of new observational techniques and forms of notation aimed at greater precision went hand in hand with the political project of developing a rational form of dressage and claims about realistic representations of equine gait in painting. A more careful reconstruction of these contexts reveals the extent to which Marey's physiology of locomotion represented not only a continuation of the Webers' experimental approach but also a recombination of practices developed in the military and in hippology.

OBSERVING HORSES IN MOTION: THE STUDY OF THE "ANIMAL MACHINE" IN CIRCUS, ARMY, AND PAINTING

The correct classification of horse gaits was a perennial topic of debate in nineteenth-century Western societies. Initially symbolizing high status through their association with aristocratic rulers and their military exploits, by the nineteenth century, horses had become not only a new status symbol for the bourgeoisie but also an integral part of the civil and military life of industrial societies.[7] The increasing prominence of horse anatomy and physiology and the passion invested in hippology as a new scientific branch within French culture were inextricably connected to debates about proper techniques of dressage. Fervent discussions, whose triumphant resolution is usually attributed to the photographic inventions of Marey and Eadweard Muybridge in the 1870s, were held about the correct representation of horses in motion.[8] To a large extent, these discussions were shaped by a number of controversies between two major schools of dressage that had divided French society since the 1830s. One main camp favored the teachings of the riding master François Baucher, the other the Vicomte d'Aure, master of the grand stable for Louis XVIII and Charles X at Versailles. The followers of Baucher advocated a method in which every individual body part of the horse that in any way resisted the control of the rider was to be subjugated systematically to the will of the rider through physical intervention. The d'Aure camp scored a victory when the cavalry training school at Saumur, which became the most prestigious equestrian school in France after the closing of the school at Versailles in 1830, made d'Aure their chief trainer in the 1840s, rejecting Baucher's method. Baucher's method, however, enjoyed great popularity in Parisian society, and France remained divided into two camps for several decades (see fig. 4.1).

    In this dispute over how to train horses, a key role was played by one of the most popular institutions of the nineteenth century, which Baucher used to make practical demonstrations of his method: the circus.[9] The circus had established itself in the late eighteenth century as a new forum for entertainment, but it also served for the demonstration of practical horsemanship. The Parisian Cirque Olympique, for example, operated by the Franconi family, presented realistic reenactments of historical and contemporary battles and can thus be considered a genre of historical theater.[10] But it also served a pedagogical function with regard to the transfer of practical knowledge of military horsemanship and provided a forum for exhibitions of equestrian vaulting (fig. 4.2). The circus ring, as a closed and controllable site, remained the preferred experimental field for the adherents of Baucher's methods in the French military after these fell into discredit. As noted, the stated aim was to subjugate and tame the horse, and this was to be done according to the laws of anatomy and mechanics. This allowed adherents to fend off accusations of subjecting the animals

Figure 4.1 François Baucher riding his horse Partisan. From Baron de Vaux, *Les Ecoles de cavalerie, Versailles, l'Ecole militaire, l'Ecole de Saint-Germain, Saint-Cyr, Saumur: étude des méthodes d'équitation des grands maîtres de l'époque* [. . .] (1896).

to an excessive use of spurs and whip, denounced by d'Aure and his followers as *boucherie* (the all-too-obvious play on "butchery" alluded both to Baucher's name and the fact that his father was allegedly a butcher).

The turn to the study of "animal mechanics" enabled a redefinition of Baucher's original maxims, which derived from the *haute école*, or "high school" of dressage, wherein "the rider acts with all of his strength on the horse, physically and morally. Through the difficult exercises to which he subjects it, he perfects its suppleness and its equilibrium. Through the

Fig. 174. — Pas de Deux.   Fig. 175. — Pas de Deux.

Reproductions d'après LE CIRQUE OLYMPIQUE ou les Exercices des Chevaux, dressés par MM. FRANCONI.

Fig. 176 et 177.— En Liberté.   Fig. 178. — Le Pas du Châle.

Figure 4.2 Equestrian vaulting at the Cirque Olympique in Paris. From Baron de Vaux, *Ecuyers et écuyères: Histoire des cirques d'Europe (1680–1891)* (1893).

continuity of his efforts, he shows the horse the extent of his influence over it and the extent to which he is its master."[11] The new approach to equestrian discipline, with its assertion of scientific rigor (as represented in the work of Charles Raabe and his followers),[12] made Baucher into the founder of hippology as a science with its own technical vocabulary.[13] According to its principles, rider and horse were in continual communication within which the "tact" and "intuition" of both parties were to be replaced by rational instruction to the greatest possible degree.

The process of educating a *cheval savant* ("learned horse," often referred to in this literature as the sagacious pupil of the horse trainer) followed the maxims of a pedagogy that, like gymnastics for children and cadets, aimed to achieve rational control of the "animal machine."[14] The goal of this training was the complete subjugation of the horse by the rider, who, as the "absolute master," was to expunge the will of the animal. This total mastery was stressed by Raabe, a cavalry captain who was also one of the most active and polemic exponents of this approach. And yet it could only be reached by respecting the limits "resulting from the animal machine: the horse trainer must avoid misusing his power by properly managing the energies involved."[15] Thus, rational dressage set itself the task of specifying the "means" by which "the horse's will can be replaced by the human will, making the animal easy to control, but under the express condition that the mechanical laws governing the horse's regular gaits never be violated."[16]

The study of equine gait viewed horse bodies as machines that functioned according to their own laws, and it was part of a pedagogical project that aimed to regulate the movement of human bodies.[17] As demonstrated in treatises by Raabe and other equestrians who set out to build the new science of hippology, the call for an *analysis* of the moving "animal machine" was raised long before instantaneous photography entered the realm of animal locomotion.[18] Via the reception in France of the Webers' work, new tools had become available: new practices for observing and representing sequences of movement; an exclusively physiological-anatomical vocabulary; and a mathematical theory of walking, jumping, and running. All of this prepared the ground for an analysis of animal bodies in motion. In the case of equine studies such as those conducted by Raabe and other horsemen, one can observe an eclectic reception of these physiological studies. One element in particular made its way into military research: the theoretical axiom of natural (two-legged) locomotion as a pendular movement, which the Weber brothers had formulated on the basis of a series of experiments (see chap. 3). After the fundamental criticism leveled by Duchenne, French physiologists considered the Weberian theory refuted, but the basic formula of pendular movement and the use of serial representation lived on.

In Raabe's "theory of the six periods," equine gait is represented as the alternating work of two oscillating pendulums, with one staying on the ground while the other is swinging. A foot in the air is conceived of as a normal pendulum, and a foot on the ground is viewed as an inverted pendulum. The amplitude of each pendulum consists of three periods (*lever*, *soutien*, and *poser* for the first pendulum, and *commencement*, *milieu*, and *fin de l'appui* for the second) for a total of six. Since both pendulums swing around the same center of gravity, the amplitude of the first, normal pendulum is twice that of the second, inverted pendulum (fig. 4.3).

A number of Baucher's adherents also used stroboscopic discs, devices that would later appear on the market under the name "zoetrope." In an appendix to a defense of Baucher's method, one Lieutenant Wachter, for example, demonstrated the correct positions of the gallop with the help of a "phenakistiscope," a spinning cardboard disc that functioned according to the same principle as the devices used by the Webers. Sequential depictions of horse gaits in silhouette appeared in a number of other publications as well (figs. 4.4, 4.5).

Up to this point, most hippologists of the era had developed their theories of natural equine gait based on the *plan de terre*, or "ground plan,"

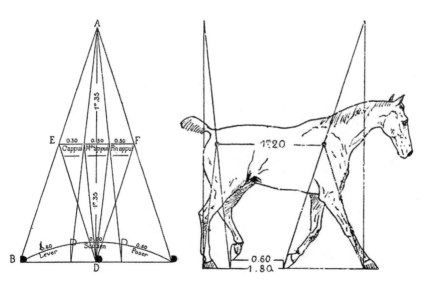

Figure 4.3 Charles Raabe's theory of the six periods.
From Étienne Barroil, *L'art équestre* (1887).

Figure 4.4 Phenakistiscope from Louis-Rupert Wachter, *Aperçus équestres* (1862).

Figure 4.5 Sequential depictions of equine gait in Gabriel Colin,
Traité de physiologie comparée des animaux (1871 [1856]).

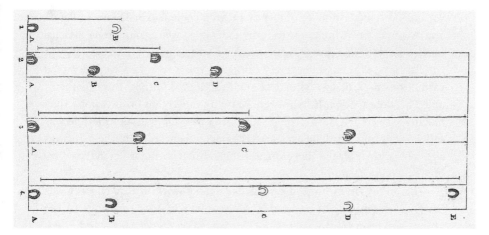

Figure 4.6 "Ground plan" in Gabriel Colin, *Traité de physiologie comparée des animaux* (1871 [1856]).

derived from the pattern of traces left by a horse's hooves on a sand surface. In the opinion of these observers, these patterns of hoofprints were the sole empirical material—and the only *sufficient* material—for analyzing equine gait, and they were translated into a relatively simple graphic notation (fig. 4.6).[19] The epistemic primacy of the hoofprint should come as no surprise: the short moment of contact between hoof and ground provided the only material evidence of an action that otherwise defied precise observation.

These practices, on which rational horse training was to be founded, went hand in hand with the question of the proper visual depiction of these animals. Thus, knowledge of the laws of animal mechanics was, for Baucher's adherents, identical with the correct visual representation of the horse: "The riding master must know the horse so well in its structure and in its organism that in every moment, with every step, in every movement he is able to visualize it both in its totality and in the function of its individual mechanisms, no matter how hidden they may be."[20] This implies that rational control of the horse is only possible if the rider is capable of continually visualizing the animal *in actu*. The most conspicuous symptom of this line of thinking was an area of specialization that developed in art criticism in which military horsemen and veterinarians spoke out as experts on questions regarding the correct representation of horses in painting.[21] One of the major authorities in this field was Lieutenant Colonel Emile Duhousset, who had made observations of various horse breeds on a number of anthropological expeditions in Persia and Algeria, and who, beginning in the 1870s, wrote numerous articles for the *Gazette des Beaux-Arts* on the correct depiction of equine gait in the fine arts. Duhousset severely criticized the way horses were depicted in motion in paintings by artists such as Horace Vernet, Rosa Bonheur, and Théodore Géricault, who were drawing mostly on observations they had made at French riding schools and circuses.[22] Duhousset's method was rather straightforward: he extracted the horses depicted from the paintings as a whole, then placed these images side by side with corrected drawings of his own creation (fig. 4.7)

The only painter praised for his exemplary work in this context was Ernest Meissonier.[23] In particular, Meissonier's history paintings in the 1860s (*Napoléon III à Solférino*, *La Campagne de France, 1814*, and *Friedland, 1807*) established him as one of the most respected equestrian painters of his era (fig. 4.8). In the judgment of Théophile Gautier, Meissonier

Figure 4.7 Corrected image by Théodore Géricault, in Emile Duhousset, "Le cheval dans l'art," *Gazette des beaux-arts* (1884).

Figure 4.8 Ernest Meissonier, *Friedland, 1807* (1875).
Metropolitan Museum of Art, New York.

had outstripped all previous achievements in the field, depicting his
horses with a "hippological science, an exactitude of movement, a cer-
titude of positions, a variety of coats, a sentiment for race of which we
know no other example."[24] It is notable that Gautier's register is not one
of aesthetic contemplation but rather one of epistemic competition and
conquest. Meissonier's horse paintings surpassed those of the seventeenth
and late eighteenth centuries because of his superior "scientific" rigor.[25]
Gautier's appraisal of Meissonier's "science" should not be understood
as merely metaphorical. According to Meissonier's biographer Octave
Gréard, the artist considered himself a scientist for whom the disciplined
observation of nature amounted to a serious military maneuver: "Being
a painter means being habituated in one's vocation to a rigorous logic, to

finding out the how and the why, and to going back from the effects to the causes. Nature only reveals its secret to those who put it under pressure. It is not enough to gaze at it with admiration; one must coerce it. 'I am naive, but at the same time I am like a drill that penetrates the things through and through.'"[26]

Meissonier was himself a passionate equestrian and had spent many years closely observing horses. In his attempt to achieve the greatest possible degree of "truth" in his art, he clothed the models for his battle paintings in faithful reconstructions of the original uniforms and worked in the outdoors. His demand that the object of representation be studied in natural surroundings corresponded with his practice of a style of academic realism that he carried to extremes toward the end of his career. The models for the series of paintings *La Campagne de France, 1814*, for instance, were obliged to stand in the freezing winter weather for hours until the desired effect for this *tableau vivant* had been achieved.[27] To a certain extent, Meissonier's history paintings appear to be a late, highbrow manifestation of the intense interest in restaging historical battles at the circus, as in the highly popular genre of the "hippodrama," which reenacted military engagements using huge numbers of extras and animals.[28]

Meissonier seems to have conducted most of his observations within a rather peculiar setup at his house in Poissy:

> In his garden he set up a long course, and parallel to it a little train track. While one of his servants led a horse along the course, the painter, sitting on a small carriage right beside the animal, was propelled at the same speed by two men. He observed every movement minutely and practiced capturing the muscular displacements of the walking animal over the course of many hours sketching everything with rapidly drawn lines. After the work of analysis had been finished, he went on to consider the movements as a whole.[29]

Exactly when Meissonier set up this arrangement for observing horses in his garden is unknown, but it must have been at least by the early 1860s, when he increasingly turned his attention to historical battle painting. From his "projet de petit chemin de fer pictural" (project of a little painting railroad), as one contemporary satire derisively referred to it,[30] one might infer that Meissonier was not so committed to the representations preferred by hippologists: the hoofprint images and *plans de terre*. His setup clearly made use of the railway as a symbol of modernity, but he articulated it in such a way, composing it from domestic servants and machines, that it only served to support the art of the painter, comfortably cushioned in his little carriage. Accordingly, the correct depiction of the horse's gait was exclusively a product of what the observer was able to glimpse and record in his sketches as he was propelled along.

THE PHYSIOLOGICAL CIRCUS: ÉTIENNE-JULES MAREY'S GRAPHICAL PHYSIOLOGY OF LOCOMOTION

In a lecture delivered in 1878 to the Association française pour l'avancement des sciences in Paris, Étienne-Jules Marey introduced the newest devices of his "graphical physiology," which he used in his experiments on "animated motors": those "obedient helpers whose power and speed is used by man according to his whim, who live in his immediate vicinity, accompanying him in his work and pleasure."[31] The "helpful living machines" he describes here were none other than animals, and specifically horses. The lecture itself was divided into two parts. In the first, Marey outlined the most rational possible use of the horse in modern transportation and in the military. In the second, he addressed the correct representation of horse gaits in hippological treatises and works of art. Marey declared that with the aid of his "graphical method" he was revolutionizing both of these areas. It was an assertion he had also made in a book he published the same year entitled *La méthode graphique*, and it had far-reaching consequences. It represented nothing less than an endeavour to free natural phenomena from a conventional system of representation and let them speak in their own language, "the language of the phenomena themselves."[32] In a historical excursus Marey claimed that this language was derived not from alphabetic writing but from "natural graphics, which in all ages and among all peoples have depicted objects in the same way, and which allow us to trace the symbols of a vanished civilization on the steles of Egypt. This graphical method of depiction would represent the truly universal language if it were applied to both the depiction of ideas and the figuration of objects."[33] According to Marey, there was only one way to exhume this universal graphical language: by mounting self-registering recording devices on human or animal bodies and translating their movements into a graph. His declared aim was not only to convey phenomena through their natural, graphical form of expression but also "to subject the largest possible number of investigations to a single method."[34]

Marey's project of implementing the graphical method as both a hegemonic form of expression and a universal research instrument was founded on a special economy in which the concentration of data, the two-dimensional compression of spatial-temporal processes, and the reduction of phenomena to mechanical factors were considered paramount for achieving "precision." Marey borrowed his method from the German experimental physiology of the 1840s and 1850s, where this style of graphical representation was first introduced by scientists such as Hermann von Helmholtz, Carl Ludwig, and Karl Vierordt.[35] The devices that Marey initially concentrated on perfecting were meant to establish a new form of producing evidence that was both time saving and compelling so as to preclude all further discussion: "if we draw the graph of the phenomena of which we are seeking to obtain knowledge and that we want to compare with one another, then we are proceeding in the manner of the geometrician, whose demonstrations are self-evident."[36] For Marey's adherents, the will to precision and the elimination of every discursively proposed critique represented the best preconditions for controlling the object of investigation.

In order to secure such control over moving bodies, Marey initially made use of an obvious strategy for physiologists of this era: he built a laboratory in which, using his new devices, the phenomena of human and animal movement would be rendered in the form of "inscriptions."[37] During the 1870s Marey developed the first complex experimental setup for the systematic and comparative investigation of human and animal locomotion. The results of these first experiments on "human locomotion" were reported in a variety of publications and summarized in the second major section of *La machine animale* (1873), whose illustrations had already acquired an emblematic status.[38] Less well known, however, is the original study by Gaston Carlet, who conducted the initial studies together with Marey in his first laboratory at the Collège de France.[39]

Carlet subdivided the prehistory of experimental locomotion physiology into three ages: a metaphysical age that began with Aristotle, an anatomical one stretching from Galen to Gassendi, and a mechanical age heralded by Borelli's *De motu animalium*. He harshly criticized the "fictitious precision" of the Webers' experiments and followed Duchenne and his electrophysiological investigations of neurological walking disorders, by then an authoritative reference point. For Duchenne, who began by observing pathological cases, gait was not the mere result of a physical force but rather "the result of admirable muscular combination" with antagonist muscles operating in harmony to keep the human body upright.[40] Carlet presented his own work on "natural gait," conceived under Marey's supervision, as complementary to Duchenne's. Like the Webers, Carlet dispensed with a definition of natural gait and insisted on the particularity of individual cases: "This work is devoted to the natural gait (everybody knows what this is). I do not claim to formulate *general* laws for those phenomena, which present so many *particular* cases to be studied."[41]

Although Carlet initially investigated human locomotion only, drawing on self-experimentation, from the very beginning Marey was interested in the comparative investigation of various "locomotion systems."[42] In marked contrast to the Weber brothers, who had restricted their investigation to human subjects, the French physiologist retained Borelli's approach to studying the gait of bipeds and quadrupeds as exemplified in humans and horses, respectively. Bipedal locomotion was considered "fundamental" to the investigation of quadrupedal locomotion:

> Dugès has compared four-legged locomotion to two men walking one in front of the other. . . . At the circus or at a masquerade, everyone has seen more or less fantastic simulacra of animals whose legs are made by two human beings whose bodies are hidden in a horse costume. This grotesque imitation takes on an astounding plausibility when the movements of the two walkers are well coordinated enough to reproduce the rhythm of a real four-legged animal.[43]

As this passage clearly indicates, Marey's first steps in locomotion physiology were also formulated based on observations of circus artists and animals. Moreover, this conspicuously placed reference provides information about the specific nature of his own experimental setup, which Carlet used to conduct his study of human locomotion. While Marey tested his new apparatus on horses at the famous riding school of Jules Pellier Jr. (a Baucher adherent) on the Champs-Elysées, the experimental arrangement at the Collège de France combined a circular path with a cylinder placed on a table on which a self-registering apparatus recorded the movements of the test person (figs. 4.9, 4.10).[44]

The higher precision of the graphical method vis-à-vis older or contemporary forms of measuring locomotion was demonstrated through an act of displacement in which almost all of the observational actions were transferred from the human senses to a single apparatus. To measure the step, Carlet constructed his own "experimental shoes," dynamometers in the form of sandals. In parallel, and following the same principle, Marey developed a prepared hoof that was used to record equine locomotion. These shoes and hooves produced no footprints; instead they were connected to a cylinder on which a stylus recorded the walker's movement in the form of

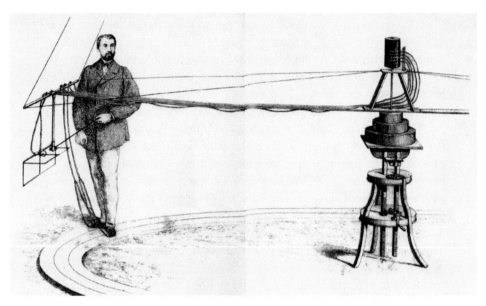

Figure 4.9 Experimental setup of Gaston Carlet for the study of human gait at Marey's laboratory at the Collège de France (1872).

Figure 4.10 E. Valton, *Horse and Rider with the Apparatus of the Graphical Method* (1874). Musée Marey, Beune.

a graph. Marey and Carlet asserted that in their graphs they could provide more insight than was possible using approaches such as measuring the distance between footprints, for their graphs also depicted the spatial and temporal aspects of movement.[45] Marey's self-recording devices, operating within a new, supposedly self-referential experimental system, claimed to produce evidence by replacing all other representations of movement.

This new experimental system was realized at Marey's Station physiologique, an open-air research site whose construction and maintenance were made possible by the close relationship between Marey's locomotion studies and the interests of both the French army and the gymnastics movement. By 1880 Marey had established contact with Georges Demenÿ, the founder of the Cercle de gymnastique rationnelle, who soon became his major collaborator at the Station physiologique. In a heated debate in parliament, the French minister of education, Jules Ferry, had supported the financing of this novel type of laboratory by stressing the practical usefulness of locomotion physiology. As he argued, soldiers could at last get shoes to reduce fatigue and the performance of horses could be optimized. The debate among the politicians clearly reveals that the experimental approach advocated by Marey, who had been appointed to succeed Claude Bernard at the Collège de France but who still completely rejected animal vivisection, appeared to be novel and unfamiliar.[46]

When Marey presented the structure and aims of the new site in an 1883 article in *La Nature*, he emphasized its healthy character by distancing it dramatically from the laboratories of experimental physiology: "sad, dingy and unhealthy rooms where researchers damn themselves to live in the mere hope that they may someday discover the characteristics of cell structure and the functions of the bodily organs."[47] In fact, the Station physiologique, set up in the outdoors, was hardly comparable to any heretofore existing laboratory: it was an "experimental field" that consisted of a combination of animal farm, circus, drilling ground, photo studio, and machine museum.[48] For his locomotion studies, Marey created a five-hundred-meter-long, perfectly level circular track with the inner lane reserved for horses and the outer one for people (fig. 4.11). The course was ringed by telegraph posts, which conducted the signals generated by human and animal subjects to the main building. A stand at the center of the track regulated gait rhythms by means of a mechanical drum, also connected to the main building. There was also a little train of the sort used by Meissonier, but constructed quite differently. Where Meissonier had a cushioned sofa on which a prominently positioned observer was borne forth by servants, Marey's rails carried a moving darkroom in which

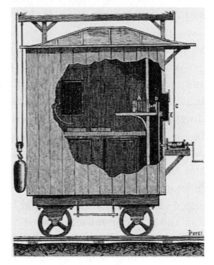

Figure 4.11 Station physiologique (*top*) and darkroom on rails (*bottom*). *La Nature*, 1883.

the scientist leading the experiment was concealed. Facing the darkroom, on the other side of the track, a black curtain was mounted, serving as a backdrop for taking instantaneous photographs of the test animals and humans as they moved past. The technique was called chronophotography, and Marey first learned of it through the new photographs of horses by Eadweard Muybridge.[49] Marey's use of chronophotography demonstratively reversed the relationship between observer and object. While Meissonier's setup made the observer more present—more visible and closer to the object of observation—at the Station physiologique, observation was an anonymous and almost completely mechanized procedure conducted from a distance. Marey wrote, "Generally it is advantageous to position the photographic apparatus relatively far away from the screen, approximately forty meters. From this distance the angle at which the test subject is being photographed changes little during the time it takes to pass the black backdrop. Red windows, through which the experimenter can follow the movements being studied, are mounted on the darkroom. A speaking tube allows the experimenter to give instructions regarding the execution of various actions."[50] Thus, the experimenter operating the camera was invisible to the test subject; only a voice was present, giving directions through the speaking tube.

In view of the tireless self-promotion that Marey undertook from the time of the Station's opening onward, it should come as no surprise that this new type of physiological laboratory became a symbol for a modernism that endeavored to mechanize all of life down to its most elementary functions.[51] This emblematic status resulted to a considerable degree from Marey's rhetoric, which left no doubt as to the frictionless operation of this almost completely automated system. As a consequence, the Station's experimental practices remained largely unaddressed, and the controversial reception of Marey's new forms of representation generally faded into the background.[52]

To a certain extent, the ideal of the automatic, anonymous production of graphical representations of locomotion explains the Station's special structure and mode of operation. Marey directed the experiments from Naples, where he spent six months of the year. He assigned the actual execution of the experiments for the most part to Demenÿ, whom he employed as an assistant (*préparateur*) and who generally oversaw the facility's finances and its cooperation with the army and other institutions. With time, Demenÿ, whom Marey initially referred to as "his workman,"[53] advanced to become the coauthor of the studies conducted at the Station, which nonetheless were generally presented by Marey personally at the meetings of the Académie des sciences and the Académie de médecine. As in the case of Gaston Carlet, who in the meantime had left Paris for a professorship in natural history in Grenoble, Demenÿ's work was assimilated into Marey's project. For many years, the practical aim of establishing a "rational gymnastics" molded experimental practice at the Station. Many of the test persons had well-trained bodies: they were athletes, and also students at the École normale militaire de gymnastique de Joinville, all of them "virtuosi" of locomotion who were to serve as models for a new, scientifically oriented model of physical education.[54]

Although Marey always presented the results of the experiments he conducted with Demenÿ as unshakable fact, both his methods and his theoretical conclusions gave rise to several major controversies. A dispute within the Académie de médecine arose as early as 1878 on the occasion of the publication of *La méthode graphique*. Gabriel Colin of the École vétérinaire d'Alfort called Marey's graphical method a "simple translation of well-known facts" and could not see that it revealed anything more about equine gait than could already be observed with the naked eye.[55] The veterinarian made the argument often leveled against Marey's graphs that their relevance was limited to the laboratory and that they were unsuitable or even dangerous for clinical work since they distracted clinicians from the direct observation of the body and were not universally readable.[56] As other episodes document, clinical medicine remained difficult terrain for the graphical method.[57]

But even among physiologists, who initially greeted Marey's locomotion studies with enthusiasm, resistance began to develop. Not long after the opening of the Station physiologique, Félix Giraud-Teulon, by now considered a leading French authority on locomotion, declared the definition of jumping that had developed out of the experiments performed there to be "physiologically impossible" and the graphs to be seriously flawed. In order to refute what he regarded as a false theory, he also produced his own graphs. A dispute that began with contention over a detail soon widened into a fundamental controversy regarding Marey's research practice at the Station. The abilities of the "new machines, so delicate and precise" (as Giraud-Teulon ironically put it) were viewed with increasing suspicion.[58] Marey sought to counter the ongoing criticism through the production of new devices and graphs. He invited Giraud-Teulon to visit the Station and step into the role of the experimental subject in order to perform the gymnastic movements himself. Still, the recordings left Giraud-Teulon skeptical about the validity of the theoretical conclusions. So Marey and Demenÿ fashioned an artificial contrivance made of lead to be attached to the foot, which simulated the movements in question in accordance with Marey's theory during a demonstration before the members of the Acdémie de médecine. As he had done after Colin's criticism, Marey ended the controversy by simply declaring all further discussion of his graphs and apparatus to be spurious and refusing to provide further answers.

Both physicians and locomotion physiologists thus proved difficult allies for Marey's project. An alliance with the military, however, seemed
more promising. By the 1870s the graphical method had attracted the
interest of military scientists, but Marey's instruments were often considered too costly or too difficult to handle.[59] The overall design of the
station and the conception of its locomotion experiments as an instrument of control facilitated its cooperation with the military, which assumed a dominant role in the studies conducted there. Marey had established contact with several of Baucher's adherents, who, in the aftermath
of Charles Raabe's work, were devoted to developing a dressage method
that was based on purely physiological principles. As is made clear in *La
machine animale*, the relationship between the graphical method and
the forms of observation practiced by the hippologists was marked by an
asymmetry: the new representations produced by Marey's apparatus were
presented as the final solution to the confusion surrounding the contradictory classifications of equine gait. According to Marey, individual
authors had defined the normal stride "arbitrarily"—that is, "according
to theoretical points of view"[60]—in no small part because their empirical foundation—the *plans de terre*—did not enable them to reach a reliable classification of the individual gaits based on locomotion physiology
(fig. 4.12).

Thus Marey initially resorted to a form of graphical representation
that combined the image of the traces left by the horse's hooves with a
notation for writing down the rhythm of the animal's stride. Eventually
the notation displaced the image of the traces left on the ground, because
the notation optically suggested a hoofprint mark expanded to include the
dimension of duration. This displacement is made clear in *La machine
animale*, where Marey juxtaposes his notations of gait with Duhousset's
drawings of horses in motion, marking points of correspondence with a
white dot.[61] Although Marey's gait notation displays a certain similarity
to musical notation, it is actually quite distinct from it. Superficially, it
resembles musical notation in that it is also a graphical representation
of sounds (in this case, the sound of steps). The major problem for Marey
with musical notation, however, was that it used too many different symbols (primarily for duration and rests) that were purely artificial conventions. His ideal notation, in contrast, would represent "the duration (of a
note) through the length of a symbol and the intensity of the tone through
the thickness of the symbol." In Marey's view, the notation of the hippologists was such a form.[62] For this reason he integrated it into his graphical representations of both human and animal locomotion—with the key

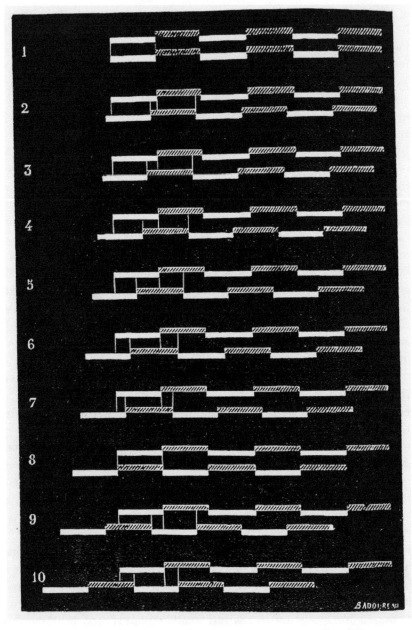

Figure 4.12 "Synoptic table of notations of equine gaits, according to the classical authors." In Étienne-Jules Marey, *La machine animale* (1873).

difference that in walking or running, test subjects notated the "music" of their stride themselves (initially as a graph, later translated into a two-lined notational system).

In view of the controversies surrounding Marey's new graphical representations of locomotion, the obvious question is whether his new approach was able to prevail and render the *plans de terre* of the hippologists obsolete. Did the graphical method revolutionize the field? Raabe, the cavalry captain, who corresponded with Marey in the late 1870s and supplied horses for the Station physiologique, stuck to his theory of the six periods and his own system of notation, categorically rejecting the need for any different form of representation.[63] However, another military horseman, Jules Lenoble Du Teil, who initially saw himself as Raabe's student, soon revised the captain's theory of the gallop after testing it using Marey's apparatus. It is hardly a coincidence that Lenoble Du Teil was the only hippologist whose work Marey praised in *La méthode graphique*. At least with regard to the rhetoric that he used to sweep aside the work of all the theoreticians before him, Lenoble Du Teil seems to have learned readily from Marey. His comment at the beginning of his own study on equine locomotion is illustrative: "I do not wish to discuss the opinions of various authors here, because the experimental resources put at my disposal by Professor Marey allow me to treat the subject with a mathematical exactitude that answers the question once and for all, thus repudiating any contradictory opinion."[64] In practice, however, the situation was not so clear cut. In contrast to Marey, who replaced the *plan de terre* and other combined forms of notation with a single graph, Lenoble Du Teil continued to use the *plan de terre*, complementing it with graphs that represented time intervals. The approach was modeled on a method of representing railway lines invented by the engineer Ibry (fig. 4.13).[65] Lenoble Du Teil's illustrations did in fact fulfil the method's definition of a movement graph on a *theoretical* level, since they included both dimensions of movement: space (in the hoofprint image) and time (in the graph). And yet these two dimensions were not represented by a single graph, as with Marey; rather, the representation of the spatial traces of movement (the hoofprints) was simply augmented by Ibry's graphs. This additive technique, which retained the older forms of representation in combined form while avoiding the decisive point—translation by the apparatus—was, according to Marey, unable to indicate the "real sequences of movement."[66]

The collaboration between Raabe's students and Marey and Demenÿ at the Station physiologique also failed to produce any lasting consensus around the correct understanding of the graphical method, and soon

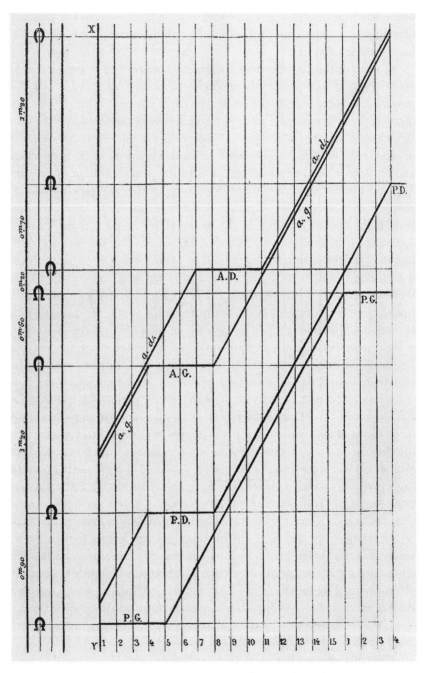

Figure 4.13 Lenoble Du Teil, ground plan of the equine gallop (1877).
From Étienne-Jules Marey, *La méthode graphique* (1878).

thereafter, the latest development, chronophotography. In 1889, shortly before Raabe's death, one of his students, Étienne Barroil, published the book
L'art équestre, bringing together the captain's life's work. It portrayed
Raabe as a brilliant observer who had anticipated many of Marey's findings and used his *plans de terre* to solve successfully many problems for
which the graphical method was doomed to fail. The book declared the
"utilization of hoofprints from the ground" to be an "indispensable complement to Marey's apparatus." Most of all, the author criticized Marey's
initial method of placing his notation under the horses' feet, thus simulating an image of the hoofprints:

> Until now the graphical notations have been used incorrectly. It has
> been considered sufficient to draw them under the feet of the horse,
> which are depicted above it in a faulty manner. This unquestionably
> proves that the graphical method alone is not sufficient for expressing
> the position of the individual limbs in the air and even on the ground.
> Thus these notations must be complemented, as has already been ex-
> plained, by the knowledge of the *plan de terre*. . . . Marey understood so
> well that the graphical method alone was insufficient that he soon re-
> turned to the analysis of the gaits, this time using chronophotography.[67]

Marey's graphical physiology appears here as a transitional approach superseded by chronophotography, although the knowledge produced by both
approaches is clearly considered inferior to the evidential status of the
plans de terre. Another of Raabe's prominent students, Henri Bonnal—the
author of several works of military history and tactics who taught at both
the École normale militaire de gymnastique and at the École supérieure de
guerre and had spent several years conducting experiments with Demenÿ
and Marey at the Station physiologique—came to the ultimate conclusion
that the photographs "certainly give rise to great astonishment, but they
can hardly serve as an instructional aid. In looking at the photographs of
horses in motion, only those who have themselves studied the mechanism
of the gaits can understand what they show, which has been captured, so
to speak, in flight, and can identify each gait's deficiencies, its advantages,
or even the position that it occupies in a series of different gaits or in the
transition between one gait and the next."[68]

The hoofprint also dominated the polemic controversy in which Lenoble Du Teil engaged Raabe's students concerning his theories. Although
he was quick to reject their notation, he was also forced to admit that his
own notations did not agree with Marey's. These differences were a result

of the ground on which the horses had left their prints: the damp beach at Deauville, over which Lenoble Du Teil had led his animals yielded different traces than the Station physiologique's circular track. Significantly, at no time did the controversy center around the most important point for Marey, namely, the use of the apparatus itself. In the end, the discussion revolved around the question of what practical knowledge of horsemanship could legitimately be translated into theory, whereby the reference to the graphical method constituted merely a gesture toward the authority of science. Accordingly, Lenoble Du Teil summarized his critique as follows: "We must come to the conclusion that the authors of *L'art équestre* were either not able to read the notation or have interpreted it under the influence of theories for which they were so filled with enthusiasm that it has obscured the clarity of vision necessary for a proper understanding of the graphs into which the facts have been translated."[69] Thus, ultimately, the controversies among his military allies ended more or less in a rejection of Marey's self-recording physiology. Just when an indisputable truth, one to silence all critics, was to be led into the arena, a new hermeneutics came to the fore.

CLINICAL TRACEOLOGIES AND PATHOLOGIES OF GAIT

During the same period that Marey was developing his new physiology of locomotion at the Collège de France, a number of medical professionals—mostly orthopedic surgeons and neurologists—were also working on recording gait, seeking to open new avenues in the diagnostics of pathologies of walking. Most of them preferred to investigate the naked foot and declined to use any kind of apparatus attached to the patient's body. The footprints that had been used in some of the physiological investigations of the 1840s returned in this new clinical traceology as it developed within various medical disciplines in the 1870s.[70]

The use of a *technique d'empreinte* (imprint technique) to fix footprints was first used in French medicine to detect cases of pes valgus, or flat feet. According to the electrophysiological investigations of Duchenne, which we have discussed at the beginning of this chapter, the condition resulted from muscle fatigue and the effect of body weight after long periods of standing or walking.[71] The literature on hygiene and military medicine noted the occurrence of flat feet especially in soldiers, leading for calls in several countries for a medical reform of footwear.[72] In 1876, the physician Ernest Onimus (1840–1915), a disciple of Duchenne, presented to the Association française pour l'avancement des sciences a new method of diagnosing flat feet, club feet, and other plantar deformities: "In certain pathological cases, one believes based on inspection of the feet alone that a curvature exists and imagines it in an exaggerated way; but a footprint taken on black paper demonstrates that this is true in appearance only, and that the seemingly arched feet flatten as soon as they are placed under the weight of the body."[73] According to Onimus, a footprint could "record not only the form of the sole of foot in a perfect and exact way but also (and this is the most important point) the parts of the foot that are in contact with the ground."[74]

Foot-ground contact thus took material form in the ink footprint, and this enabled physicians to diagnose gait as either normal or pathological. This semiotics of traces was premised on a mechanics that recalls the Webers and their ideal human walking apparatus. Ideally, the rolling motion of the feet should occur with minimal ground contact: this "provides the best result for the muscular effort, since, as with the wheel in mechanics, the surface in contact with the ground is reduced to a minimum."[75] When it came to defining normal gait, however, this clinical reasoning used forms of representation that differed considerably from those of the Webers. Onimus printed reproductions of gait patterns derived from ballerinas or "skilled pedestrians," saying that normal walking resembled in astonishing ways the act of dancing: "The least tiring and most gracious gait is one in which the foot is never placed flat on the ground, and at the last moment of the rolling motion the outer toes are involved. . . . Thus for a brief moment one assumes the attitude of a ballerina in the sense that the entire weight of the body is carried by the toes."[76]

Onimus also presented his rather simple technique for producing footprints "as a kind of graphical method,"[77] a gesture toward his colleague Marey, whose self-recording instruments he had criticized previously in another context.[78] Subsequently, other physicians also adopted this label for the procedures they used to classify pathologies of gait.[79] In marked contrast to Marey's apparatus, which interrupted the contact between the feet and the ground in order to render the laws of natural movement visible, this neurological and orthopedic diagnostic practice required the analog representation of the foot impressed on the material surface of contact. The footprint assumed its privileged epistemic status only because physicians claimed to be able to isolate a "type" from all the individual manifestations of walking encountered in everyday life or in ethnographic or clinical observation. Without the demonstration of the virtuosic capacities of the observer, however, the diagnostic tableau of traces remained rather opaque.

Within the discipline of neurology as it became established in the
second half of the nineteenth century in Germany, France, and England,
detailed observation and analysis of the gait became an essential component of a scientific approach to diagnostics. This observational method
was demonstrated not only in the clinical lecture hall by the "examination of the *living* subject"[80] but also in richly illustrated textbooks.[81] In
exemplary fashion, the neurologist Jean-Martin Charcot presented in his
famous "Tuesday lessons" at the Salpêtrière hospital in Paris a differential diagnostics indebted to Duchenne's studies of the gait in hereditary
muscular pathologies (such as "locomotor ataxia," also called "tabes dorsalis").[82] Charcot's Tuesday lectures were designed as clinical demonstrations based on the principle that "seeing is better than reading" and "one
can learn far more from a quarter hour spent examining patients than
from studying the descriptions of their ailments in books."[83] In order to
differentiate among neurological diseases such as Parkinson's disease and
multiple sclerosis, Charcot's method combined long-term observation of
living patients with analysis of autopsy records. This approach, generally
known as the *méthode anatomo-clinique*, was enriched, augmented, and
modified over time, especially through more systematic study of the nervous diseases by Charcot and his pupils.[84]

From the late 1870s on, the Salpêtrière used a variety of self-recording
devices designed by Marey to register the bodily movements of patients.
While this partial application of the graphical method was useful in in-
vestigating some illnesses, such as for recording the tremors of multiple
sclerosis or Parkinson's disease,[85] in others, notably chorea and hysteria,
the apparatus seemed unfit to record the rapid and complicated progression of body movements. As in the case of other disease entities, Charcot's
formalization of *la grande hystérie* (great hysteria) hinged largely on the
medical art of observation wherein the trained eye and ear of the physician acted as the final arbiter.[86] In the semipublic Tuesday lessons, Charcot made repeated demonstration of the primacy of the neurologist's own
senses, especially for when it came to detecting relevant signs in the study
of gait. Thus, in a complicated case of locomotor ataxia or tabes dorsalis,
Charcot insisted on the necessity of distinguishing the gait of the alcoholic from the tabetic gait: "You already know that the ataxic throws his
legs and feet forward as he walks. The typical alcoholic, on the other hand,
bends his knees, excessively like a prancing horse. This alcoholic's gait
has been named *démarche du 'Steppeur,'* the 'stepping gait,' after the English 'stepper.'"[87] Textbook knowledge, in this case the classical descriptions by Duchenne and Romberg of the tabetic gait, was only a starting

point and subject to constant interrogation by clinical observation, which was able to reveal the great variability in walking disorders.[88] Repeatedly, Charcot presented patients to his audience whose gait presented the features of two distinct neurological diseases at once. The initial ambiguity about the correct diagnosis led to accepting a certain striking walking pathology as a combinatory form whose corroboration was then undertaken by checking other classical neurological signs (reflexes, abnormalities of speech) and also by an orthopedic inspection of the patient's feet using footprints.[89] But as other lessons show, these forms of visual evidence also had to be complemented by listening to the sounds produced by the patient's distinct walking patterns. Comparing the alcoholic and the ataxic patient, Charcot said that the former "overflexes at the knee joint and the thigh rises more than it should. As the foot hits the ground, the toes strike first and then the heel, so that you can quite distinctly hear two successive sounds." The ataxic patient, in contrast, "thrusts his leg forward in extension with almost no flexion at the knee joint; this time the foot hits the ground all at once, making only a single noise."[90]

In this case and in a series of others, diagnostic work is presented as a virtuoso performance by the physician. The physician's eyes, hands, and ears identify disease through the comparison of different gait pathologies recognized as characteristic of distinct clinical types. Charcot's clinical artistry was often described and praised by his pupils. However, it should be noted that the book version of the Tuesday lessons did not translate this knowledge into a visual medium. The transcriptions reproduced mostly the text of the lectures, illustrated only occasionally by drawings, most from Charcot's own hand.

This primacy of drawing over photography was characteristic for the observational practice that prevailed at the Salpêtrière (not only for financial reasons but also for epistemological ones linked to Charcot's conception of the clinical type).[91] Thus, it is not surprising that for registering patients' gait, footprints remained the preferred method. In his 1885 prizewinning medical dissertation on clinical and physiological studies of walking, Charcot's pupil Georges Gilles de la Tourette (1857–1904) characterizes his own technique as a *méthode des empreintes*, an "imprint method," in part to underline the extent to which it differed from the "graphical method" employed and refined by Marey and his pupils.[92] His subjects had to first soak their feet in a solution of red ink and then walk down a line drawn on a long strip of paper. The footprinted paper was then photographed and the photograph reduced to a smaller size for publication (fig. 4.14). Gilles de la Tourette notes the complete absence of

TABLEAU COMPARATIF DES DIVERS GENRES DE MARCHE

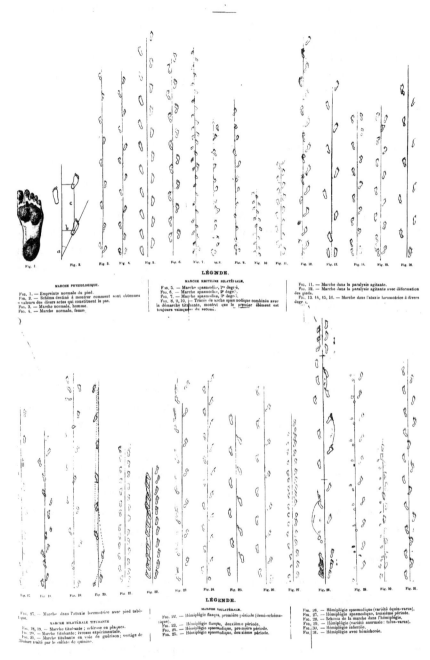

Figure 4.14 "Comparative tableau of different forms of gait," from Georges Gilles de la Tourette, *Études cliniques* (1885).

any apparatus that had to be mounted on the body as the crucial advantage of his method, which he advertised as "the only one that can be applied in the clinic."[93] In order to compare normal locomotion with pathological forms, one had only to let "normal" subjects and several patients walk across the strip and then observe their deviations from the drawn line. A setup involving shoes was rejected as unable to give an accurate representation of the act of locomotion itself. Here, Gilles de la Tourette criticized the German physiologist Hermann Vierordt (1853–1944), who had proposed a method of studying human locomotion inspired by Carlet's and Marey's initial setup.[94] Vierordt replaced the circular path used by the Frenchmen with a *Direktionslinie*, "guiding line," drawn on a strip of paper on which the subjects had to walk in an empty hall in the physiological institute of Tübingen. In order to "preserve the purest expression of the movement as it actually occurred in space,"[95] Vierordt designed a *Spritzmethode*, "spray technique," that combined Marey's graphical physiology with the traceologies of the clinic. To guarantee that the "spatial relations" of walking were recorded, subjects wore special shoes with three tubes attached containing sponges saturated with differently colored liquids, one at the heel and one on either side of the sole. The reservoir that fed the liquid to the tubes was placed partly on the back, partly on the head of the subject. As in Carlet's initial setup, the temporal relations of walking were recorded on a rotating cylinder, although in this case not through pneumatic but through electric impulses (fig. 4.15).

This rather curious method elicited the criticism of Gilles de la Tourette not just because it necessitated the use of "special, obstructive shoes, which had to render faulty the results thereby achieved,"[96] but also because it reduced the entire sole of the foot to three points, which in some pathologies would not all make contact with the ground. Although Gilles de la Tourette singled out the German physiologist Hermann Vierordt for his criticism, the neurologist clearly distanced himself from the extension of Marey's "graphical physiology" to the pathologies of movement described by Duchenne and Charcot in their differential diagnostics in such minute detail. It is significant in that context that Marey, in 1884, discussed the problems raised by Vierordt's setup, but that was mostly so that he could present his own new chronophotographic technique as a solution.[97] Shortly before Gilles de la Tourette's dissertation appeared, Georges Demenÿ was already busy developing, under Marey's guidance, a complex system of light points that was to allow the successful transfer of his experimental approach from the open-air setup of the Station physiologique to the neurological clinic.[98]

Fig. 1.

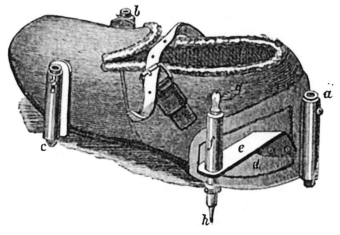

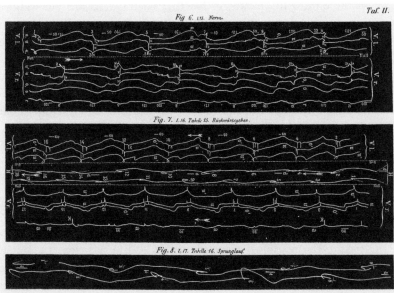

Figure 4.15 *Top*, Hermann Vierordt's experimental shoe. *Bottom*, Gait images, plate 2.
From Vierordt, *Das Gehen des Menschen in gesunden und kranken Zuständen* (1881).

Installing a laboratory in the clinic occupied Demenÿ and the surgeon Edouard Quénu of the Hôpital Beaujon for two years. The project proved difficult, especially because the quality of the chronophotographic recording technique was critically dependent on light. It involved attaching small lights to different parts of patients' bodies and having them walk in front of the chronophotographic apparatus in a dimly lit room. The lights were connected to a battery by an elastic cable that ran from the subject's back to a device fixed above the head so as not to obstruct the subject's movements (fig. 4.16).[99] This cumbersome and costly method could not compete with the simple *méthode des empreintes*, and thus it remained (like Vierordt's spraying technique) an unfruitful episode.

During the last decades of the nineteenth century, these clinical traceologies shaped the epistemic models and practices of many of the emerging human sciences. In the clinical literature, the reproduction of footprints combined with drawings or photographs of characteristic expressions of the patient's body or face became part of a physiognomic project that sought to assign specific forms of gait to places within a typology. The advocates of the *méthode des empreintes*, a method that found its way,

Figure 4.16 Experimental setup of Quénu and Demenÿ for the
study of gait pathologies using the chronophotographic method.
From Louis Gastine, *La chronophotographie* (1897).

in modified form, into criminal anthropology, were therefore less concerned with documenting the variability of individual body postures and forms of gait than with trying to make visible their typical expressions.[100]

INSTANTANEOUS PHOTOGRAPHY AND THE "OBJECTIVE AESTHETICS" OF MOVEMENT

In his 1902 study *Introduction à la figure humaine*, Paul Richer declared that "instantaneous photography has opened people's eyes. The revolution has not occurred without resistance, but today it is a fait accompli."[101] For this artist and physician, who after a long association with Charcot at the Salpêtrière now held a prominent position as professor at the École nationale supérieure des Beaux-Arts, it was evident that Marey's chronophotographic studies had fundamentally transformed artistic observation. It had "taught them [artists] to see nature better and consequently also to interpret it better."[102]

Remembering the incompatibility between Charcot's clinical art of observation in the domain of pathological gait and the mechanized recording practices Marey employed in his physiological experiments, Richer's apodictic judgment in favor of the latter may seem surprising. Richer, however, had already advocated for a "scientific aesthetics" that aimed to synthesize clinical and physiological methods of recording gait in his 1895 book *Physiologie artistique*, a collaboration with the photographer Albert Londe.[103] And even if Charcot's semiotic practices of neurological observation and Marey's chronophotographic analysis of movement in distinct, mechanically registered moments differed from each other in every respect, Charcot and Marey nonetheless shared a view that the major task of the fine arts was nothing but the faithful imitation of nature. Like Gerdy, the Webers, or the Baron Dupin (whom he cited explicitly), Marey made a fervent plea for giving artists scientific instruction to ensure that their representations of movement were "correct." Although Marey expressed a disinterest in artistic questions, he did insist that physiology could play a central role by providing artists with good raw material to draw on. In Marey's view, instantaneous photographs offered painters and sculptors a set of possible positions that they could use to bring "interesting diversion" to the all too conventional representations of movement in the arts.[104]

Richer shared this view, one that elevated the scientist to the rank of expert in the matter of the correct representation of natural physical posture. In numerous works, Richer took up the morphology of the human body in motion, declaring that its ideal expression could be found both in Greek antiquity and in the products of chronophotography. True to the practice of the neurological clinic, the "science du nu" championed by Richer called for precise observation of the naked human body, preferably that of well-trained subjects (such as athletes), for the purpose of artistic instruction. Drawings were the primary medium for capturing movement, but a close interrelationship soon emerged between these and the mechanical recordings of instantaneous photography (fig. 4.17).[105] In parallel with the efforts of Duhousset to establish a canon for the correct depiction of horses, Richer worked for many years on a canon of the human form that took its starting point in the living body, not the dead one.[106]

Despite the freedom that Richer granted the creative artist in aesthetic matters, he nonetheless assigned the scientific data produced by locomotion research a normative function. The norm they enforced lay at the origin of all art, in the "primitive art forms" of antiquity, whose scientific accuracy was revealed by Marey's chronophotographs: "this shows us with what fidelity and precision these first artists, as yet unbound by the fetters of tradition, were able to see and observe nature."[107] Scholars of antiquity also welcomed the study of the physiology of movement and incorporated its results into their own research. Taking up Marey's rapprochement between the archaeological method of basing reconstructions on a sort of "natural graphics, which in all ages and among all peoples have depicted objects in the same way,"[108] and his own graphical method for tracking body movements, the French archaeologist, art historian, and religious scholar Salomon Reinach soon began to use chronophotographs as a tool in his interpretations of antique vases and reliefs. In Marey's images of jumping men, Reinach saw evidence that certain kneeling figures in Greek antiquity represented the springing step of the Erinyes.[109] Similarly, the musicologist and composer Maurice Emmanuel took his cue from locomotion physiology when he classified images on vases and other "monuments figurés" in the Louvre in an attempt to reconstruct ancient Greek dance.[110] Those classical representations of movement were considered the most primordial expression of the pure observation of nature. Deemed entirely unspoiled by the distorting influence of artistic styles and traditions, they were often elevated by critics of cultural "decadence" to models for aesthetic and educational reforms.

Still, the authority of Marey's scientific representations in matters of art did not go uncontested even by authors who did recognize the usefulness of instantaneous photographs. In the case of Marey's first (and only) book on chronophotography to be translated into German, a critique found

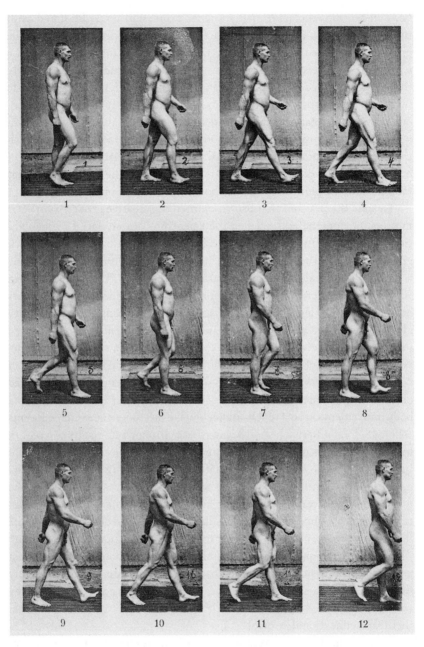

Figure 4.17 Chronophotography by Albert Londe. From
Paul Richer, *Physiologie artistique* (1895).

its way directly into the text in the form of a number of extended footnotes added by the translator, the philosopher Adolf von Heydebreck. Although Marey had refrained from making any aesthetic judgments in his 1893 book, the translator nonetheless felt obliged to add several remarks "in defense against faulty conclusions drawn in the comparison between the artwork and the instantaneous photograph."[111] Von Heydebreck was endeavoring to clarify the differences between painting and sculpture, arguing that the choice of medium played a decisive role in determining whether or not it was possible to depict the "instability" that Marey demanded in artistic renderings of movement. "Any position that the sculptor should endeavor in any way to exploit artistically must at least provide . . . the appearance of a certain, albeit momentary, stability."[112] Furthermore, the philosopher contested the notion that the unaccustomed positions and postures revealed in Marey's photographs could usefully be exploited for aesthetic ends. Such postures were, he writes, "in the highest degree disharmonious and unbeautiful to the point of ridiculousness," at least to the "unprejudiced fantasy of the natural human being, who is, after all, the only one for whom the artist works."[113]

Critical objections were leveled from other directions as well. A number of Marey's fellow scientists also disputed the relevance of instantaneous photographs for the fine arts. The physiologist Ernst Brücke, for example, who had also taught anatomy at the Akademie der Bildenden Künste in Berlin, held the opinion that artists should not take cues on moving bodies from chronophotographs. In such "realistic" attempts to copy nature, Brücke saw the main cause of an artistic decline:

> The realism now universally dominant is taken by many of our artists to mean that the praiseworthiness of their work depends on the degree to which they have faithfully copied their model. Everything, the beautiful and the ugly, is imitated to avoid becoming *conventionell*. And how dangerous this copying of a model is! . . . It is a lowly comparison, but forgive me, because it is fitting: the artist should know the faults of the human form, as the horse expert knows the faults of the horse's form. But he does not have to become monotonous, to make his forms imitate conventional themes. He can seek out beauty in its various manifestations.[114]

The lowly but fitting comparison to the horseman was not chosen by coincidence. Since the early 1880s, Brücke had taken part in discussions surrounding depictions of horses that were revealed to be flawed based on the findings of instantaneous photography. He came to adopt a completely different position from a number of his colleagues, most notably Emil du Bois-Reymond, who argued for the authority of the mechanical recording methods.[115] By contrast, Brücke wrote that "the value of perfectly passable depictions of trotting horses has in recent times been called into question on account of precise experiments, but wrongly so. . . . Presumably the artist captures the trotting horse at a certain moment: the moment that derives from his memory image. The photographic depiction, however, captures it at a random moment, and it must be a special coincidence when this moment coincides with that of the artistic depiction."[116] If one followed this line of argument, the question of which instantaneous photograph the artist should select to achieve a successful depiction no longer mattered. According to Brücke, the only possible basis for depicting locomotion was the "memory image," in which natural sensory data and the artist's aesthetic judgment merged:

> The artist does not depict an arbitrary moment of movement but rather that which would leave behind the clearest memory image in the beholder if the movement were to occur before his eyes. When he must select from moments of equal value, he selects either the one that is most characteristic for the action . . . or the one that fits best on the basis of artistic, of aesthetic considerations. Here he must make connections with memory images in the beholder that correspond with earlier visual impressions. When the impulse has been given and given properly, it is the psychical activity of the beholder that breathes life into the work of art.[117]

By formulating the problem from the perspective of the physiology of sensation, Brücke brought into the discussion the issue of perception itself, which had been consistently ignored by the locomotion physiologists. This maneuver neutralized their asymmetric definition of the relationship between the human eye and the observational apparatus inserted between human being and nature. And thus the assumption that human seeing had been fundamentally transformed through instantaneous photography now demanded more precise explication: a transformation of seeing could be understood in the sense of a displacement of attention to details or a transformation of perception or of visual habits. Stimulations of the retina entering consciousness could be described as "seeing" just as much as the visual recognition of a previously perceived object in the form of a memory image. Such problematizations, which formed the germ of a psychology and an anthropology of seeing, shifted the question of the "objectivity" of artistic depictions into another register.[118]

VISUAL ANTHROPOLOGIES AND WALKING REFORMS

At the end of the nineteenth century, chronophotography was also discovered by the first French anthropologists, who had understood their own discipline since the 1860s as an anthropometric project of detailed observation and measurement of bodies with regard to racial characteristics.[119] In the initial days of the Société d'anthropologie, Emile Duhousset provided reproductions of footprints of the Kabyle people, admiring their ability to walk very long distances without fatigue.[120] Other Société members drew on prehistoric traces and ethnographic observations to postulate the existence of a "primitive" form of walking that had been superseded by Western civilization.[121] In 1893, Léonce Manouvrier (1850–1927), a disciple of Paul Broca, described a gait he called *la marche en flexion* in which the trunk was bent over and the knees were flexed, a manner of locomotion supposedly used by human ancestors during the Quaternary period.[122] His colleague Félix Regnault (1863–1938) found the best expression of this gait in a painting of a soldier by the French academic painter Edouard Detaille, a disciple of Meissonier (see fig. 4.18):

> He does not lift the foot as much as necessary to avoid the irregularities of the terrain, the knees are strongly bent; the trunk is bent forward as much as possible. The foot detaches itself abruptly from the ground to move forward, the tibiotarsal joints are half-bent [*à demi-flexion*]: we can call this a half-flexed gait [*en demi-flexion*]. This form of gait is less unusual than we might think. Soldiers often adopt it at the end of a long loaded march. . . . Our peasants, especially the mountaineers, practice it as well; we consider it plump and ungraceful, but they progress very rapidly and much faster than the townsman. Most of the savages, especially the Negroes, also walk like this.[123]

Regnault reached his conclusions based on comparative studies he had done using Marey's chronophotographic apparatus. In 1895 he produced numerous series of photographs of Africans who had been put on display at the ethnographic exhibition at the Champs de Mars park in Paris (fig. 4.19).[124] Like the work of Paul Richer, who projected at the same time his anthropometric canon of the human body, the primary role of instantaneous photography was to provide tangible evidence of previous

Figure 4.18 Edouard Detaille, *A l'étape*. From Félix Regnault, "La
marche et le pas gymnastique militaires," *La Nature* (1893).

ethnographic observation. Juxtaposed with these archival materials documenting the "primitive" walk, the gaits of both the modern citizen and
the soldier suddenly appeared unnatural, forms of movement shaped by
aesthetic conventions. Along with Albert-Charlemagne-Oscar de Raoul, a
squadron commander who claimed to have experimented with the *marche
en flexion* in the French army since the early 1870s,[125] Regnault emerged
as a spokesman for a general reform of walking, one that sought to erase
all inefficient forms of gait. During the 1890s, the two men and their assistant Charles Comte conducted a number of experiments at the Station
physiologique that were meant to prove the superiority of their new gait.[126]

Figure 4.19 *African Children* (1900). Film by Félix Regnault
and Étienne-Jules Marey. Cinémathèque Française.

De Raoul acted as one of the principal test subjects. He was one of those "virtuosi" of running and walking who could attain 18–24 miles per hour without exhaustion. Through such extraordinary exploits, he claimed to surpass the most famous walkers and runners of his time, including Firmin Weiss, widely known and admired as "l'homme étincelle" (loosely, "the firecracker").[127]

De Raoul and Regnault drew a number of conclusions from studies undertaken by Marey and Demeny on the common gymnastic step, which had been adopted in the French army since the Second Empire via the teaching of Napoleon Laisné.[128] The fact that the body seemed to lose contact with the ground for a short instant and thus had to suffer from "successive shocks" would inevitably cause early fatigue.[129] It is perhaps not surprising that in their manual *Comment on marche* (1898), a summary of their studies with an enthusiastic preface by Marey, Regnault and de Raoul portray the Prussian goose step as the most artificial and unhealthy deformation of the natural gait. In their view, not only did goose-stepping produce the same dismal effects as the old French gymnastic step—great exhaustion and fatigue—but the physical toll it took also transformed the soldier into an "automaton," devoid of "will and independent spirit."[130] Physical training, then, was considered to be a kind of hypnosis session in which the trainer influenced the "exhausted subject" through the power of suggestion, a model that Regnault and de Raoul saw at work not only in the German army but also among the student-athletes at Cambridge.[131] However, to see their comparative overview of gait by culture and race as a mere reflection of nationalist or racist beliefs would be an oversimplification. On the one hand, their diagnosis of the physical and mental "surcharge" or strain differed little from criticisms voiced by many German physiologists and physicians who also conducted experiments into the effects of marching on respiration, circulation, temperature regulation, and nervous and muscular activity.[132] On the other hand, the final and major revelation of their comparative study, enabled by chronophotography, was that the habitual everyday gait of the French citizen was nothing less than a slightly modified form of the Prussian goose step, a form of walking both "ungracious" and unhealthy.[133] The tenor of their work was critical not only of other national styles of walking but also of Western civilization and education in general. The point was driven home by Marey, who bemoans in his preface that "we are the slaves of conventional aesthetics, both in regard to walking and to all other processes of life. This is because we have been taught since our childhood that to have a distinguished gait one must hold the chest straight, not move one's arms, and turn one's foot outward with an extended knee when putting it on the ground. Aesthetics are everywhere."[134] According to the physiologist Marey, his new scientific representations would help shake off the yoke of "ridiculous" convention imposed by the dictates of fashion once and for all: "Chronophotography, combined with the use of dynamometers, provides us with exact information about all of the acts we execute, which often reaches our consciousness in a very incomplete form. In this way it can educate our movements, allowing us to recognize the ideal perfection that we should attain and to recognize both our incorrect movements and the progress that we have made."[135]

The idea that chronophotography could ultimately transform automatically acquired habits of human locomotion was not far removed from the vision of physical training as a form of hypnotic suggestion. The epistemic and political quest for the mechanical rectification of impractical postures and gaits clearly resonated with the uses of a new psychology of suggestion in pedagogy during the last decade of the nineteenth century.[136] While both the therapeutic and military applications of the *marche en flexion* would soon be abandoned,[137] the project of a comparative anthropology of body movement based on a chronophotographic archive would survive into the next century. As early as 1900, Regnault imagined the establishment of chronophotographic collections of physical postures and typical movements of native subjects to enable the analysis of what he called their "gestural language."[138] According to this view, ethnography would become a true science only when it returned to the analytical spirit of Marey's cinematic physiology, whose task had been to decompose bodies in motion rather than to produce the illusion of movement, as was the case in the widespread and increasingly popular uses of cinema. Although this vision—of an anthropology that pieced together foreign cultures and bodies by relying exclusively on mechanical recording and purported to eliminate the subjectivity of the ethnographic observer—would find its proponents, it ultimately turned out to be a cul-de-sac.

CONCLUSION

The Centipede's Dilemma

A centipede was happy—quite!
Until a toad in fun
Said, "Pray, which leg moves after which?"
This raised her doubts to such a pitch,
She fell exhausted in the ditch
Not knowing how to run.

Anonymous, *The Centipede's Dilemma*

Looking back at more than a century of locomotion research in its most exemplary manifestations, one reaches the conclusion that this mundane activity has proven remarkably recalcitrant to being transformed into a scientific object. That is why this book has not made the attempt to cast this history as one of progressive objectification or of the consolidation of a clearly defined field of research with its special laboratories, instruments, standard methods, and definitions. And while the "science of walking," whose foundations the Weber brothers sought to lay, would always retain close ties to military and educational institutions, the new methods designed by the physiologists of locomotion did not culminate in normalized forms of walking and marching. As opposed to a teleological view of mechanization taking command,[1] this book has attempted to show how other scientific, medical, and artistic approaches led to multifarious varieties of knowledge about the human gait. These forms of knowledge seem to be better understood as an interplay between mechanical, semiotic, and poetic registers.

The predicaments of a pure mechanics of the human gait, as it was conceived by the Weber brothers and later promoted by Marey, become apparent in the last great attempt to formulate such a mechanics, under-

142

taken by the Leipzig anatomist Christian Wilhelm Braune in collaboration with the mathematician Otto Fischer. Begun in 1889, their study *Der
Gang des Menschen* was published between 1895 and 1904.[2] Although the
two scientists considered the Webers' theory of the swinging pendulum to
be an idealization of real processes, they were not convinced that it had
been actually refuted. That is why they presented their own research and
definition of human locomotion as a direct continuation of the *Mechanik
der menschlichen Gehwerkzeuge*, a text that Braune prepared for an edition of Wilhelm Weber's complete works after the latter's death in 1891.
By combining anatomical research with physiological experiments on soldiers, Braune and Fischer attempted to provide a more solid empirical basis
for the Weberian laws. Taking the *Wanderschritt* (hiking stride) as their
norm, they measured its average length and duration in experiments on
more than a hundred members of the Royal Saxon infantry. On this basis,
they claimed their results fully concordant with Marey's contemporary
studies on the physiology of locomotion.[3]

In contrast to the French physiologist, however, the two Germans
moved their subjects from the outdoors back into a dark laboratory and
loaded them up with heavy apparatuses. To light the body parts necessary
to their gait analysis, they dressed their subjects in black and attached
electric tubes to their garments. The indoor setting fixed the problem of
unwanted reflection from the sun, but the cumbersome clothing—the special suit the subjects wore was designed to protect them from the high
voltages of the induction currents—necessitated a preparation time of several hours (fig. 5.1). To enable entering the data recorded for multiple moving body parts into a coordinate system, four mobile cameras were used
to record the moving bodies from different distances. The mathematical
analysis of the data took more than ten years. At the end of his seven-
hundred-page study, replete with tables and formulas, Otto Fischer was
forced to admit that "the much-discussed pendulum theory of the Weber
brothers is false."[4] The project undertaken by Braune and Fischer, then,
illustrates again in a nutshell the central paradox of a physiology of locomotion oriented toward the ideal of homogenous, frictionless movement.
The natural functioning of the human walking apparatus could only be
grasped by an intricate technical set of experimental devices.

Although mechanical metaphors were in circulation in medicine and
the sciences around 1900, they did not provide coherent models for understanding the workings of the human walking apparatus. In the field of
neuropathology, multiple cases of walking disorders that were classified
as "hysterical" posed a challenge to physicians and scientists who tried

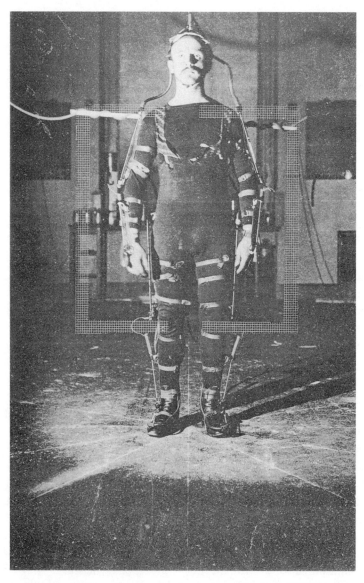

Figure 5.1 Test subject in experimental gear. Plate 1 in Wilhelm
Braune and Otto Fischer, "Der Gang des Menschen I. Theil: Versuche
am unbelasteten und belasteten Menschen" (1895).

to master them during this period by resorting to experimental and therapeutic uses of hypnotic suggestion. Freud's psychoanalysis, which emerged
in this conflict-ridden field, advances from the initial mechanistic conception of the "psychical apparatus" to an understanding of abnormal forms
of gait and motility through the exploration of psychical trauma.[5] In two
case studies in the *Studies on Hysteria* (Elisabeth v. R. and Cäcilie v. M.),
Freud discusses the onset of pain leading to gait disorders that do not
correspond to "any recognized pathological type."[6] His method of interrogation and abreaction aims directly at the legs, which begin, when the
analysis arrives at crucial points in the patient's earlier history, to "join
in" (*mitsprechen*) the patient's discourse by producing painful symptoms.
This artificially induced pain serves Freud as a "compass": "if she stopped
talking but admitted that she still had a pain, I knew that she had not told
me everything, and insisted on her continuing her story till the pain had
been talked away."[7] The leg pain cannot be described as a unified organic
symptom but is revealed to be a complex formation, composed of different
similar symptoms, resulting from specific body positions in earlier trau-
matic situations in the patient's life. Even if the seamless translation of
somatic into psychical events does not succeed completely in the case of
Elisabeth v. R., Freud does use this example to corroborate his hypothesis
that the hysterical gait disorder (like other symptoms) is created through
a process of "symbolization." The efficiency of this process is illustrated
by the case of Cäcilie v. M., who had a shooting pain in the right heel that
made walking impossible. Freud explains the pain by his patient's fear of
"find[ing] herself on a right footing" in a society where she is confronted
with strangers.[8]

It does not come as a surprise, then, that finding one's feet would soon
become a trope indicating the negative effects of physiological knowledge
about walking in a clinical context. In Italo Svevo's fictitious autobio-
graphical novel *La coscienza di Zeno* (1923), Zeno Cosini is the neurotic
son of an entrepreneur whose incessant preoccupation with medical theories and therapies only has the effect of worsening his state. In a representative episode, he learns from an old friend who is devoted to the study
of his own gait disorders that "in a brisk walk, each step takes no more
than half a second, and that during that half second no fewer than fifty-
four muscles are set in motion." The information, which is processed
instantaneously through Zeno's nervous system, does not have curative
effects: "My mind boggled and my thoughts immediately traveled to my
legs, looking for this monstrous machine. I believe I found it. Naturally
I did not find fifty-four separate devices but rather a large and complex

mechanism which lost all its orderliness as soon as I directed my attention
to it."[9] Conscious awareness of the "monstrous" walking apparatus and
its complex machinery robbed it of its ability to function and made walking, for Svevo's protagonist, "a toilsome, even slightly painful endeavour":
"this tangle of mechanisms had lost their lubricating oil and now impeded
one another in their motion."[10]

Svevo's comical episode, which highlights the difficulty of channeling
the instrumentality of walking without friction and in useful fashion into
reflexivity is a variation on a theme known by psychologists of the time
as the "centipede's dilemma."[11] In Gustav Meyrink's allegorical tale *Der
Fluch der Kröte* (1903), a toad requests a centipede to think consciously
about his own mechanism of walking, thus condemning the latter to total
immobility.[12] Already in 1889, when locomotion physiology enjoyed great
prestige, the British zoologist E. Ray Lankester had proposed to apply Muybridge's technique of instantaneous photography not just to equine gait
but also to the gait of centipedes and other complicated forms of insect
locomotion.[13] Not without irony, he cited an anonymous poem in order
to indicate the potentially "catastrophic consequences" of locomotion
physiology on the act of walking. Could the mechanical recording of gait,
cultivated by physiologists for more than two decades, happily solve the
centipede's dilemma, or was it more likely to aggravate it instead?

The problem of reflexivity in the act of self-observation, which appeared as the curse of a purely mechanical locomotion physiology, was
thus one impetus for a new orientation of the field. There were others as
well. The limits of biomechanics were not only psychological but also
mathematical in nature. A manual of physiology published in 1930 painted
a rather bleak picture of an "exact mechanics of the human walking apparatus": "The human body consists of 213 individual, more or less mobile
bones and 322 muscles, each christened with special names. Every one of
these muscles consists, on average, of a few thousand individual elements
(fibers). The muscle fibers also can change their lengths and tension according to laws whose simplest foundation has barely been elucidated so
far. It is therefore highly unlikely that the locomotive processes of such
an immensely complicated system could be described or explained with
primitive devices in a truly exact manner."[14] Even if it were possible to
conduct such an "exact" analysis for any one human, there would be the
problem of transposing the measurements made to another individual,
whose body proportions and composition differ completely. And then, according to the "holistic" trend, the states of consciousness would also have
to be taken into consideration.[15]

Concern for a holistic understanding of locomotion is also visible in the anthropological conception of walking as a "technique of the body," which Marcel Mauss introduced in a famous lecture before the Société de psychologie in 1934. According to his definition, "the body is man's first and most natural instrument. Or more accurately, not to speak of instruments, man's first and most natural technical object, and at the same time, technical means, is his body."[16] The diversity of the "techniques" of which the body is supposed to be the seat is great: Mauss' list comprises ordinary forms of walking, different marching steps to be observed in various national armies, the positions of arms and legs during walking or running, practices of eating and body hygiene, acts of sexual intercourse, and also of sleeping, swimming, and dancing. It is striking that Mauss introduces his notion by insisting on the biological foundations of the "total man" but without any direct reference to the anatomy or physiology of locomotion.[17] His crucial examples are not drawn from the scientific investigation of human motion but instead from informal observations on his own and others' body techniques. It is not an exaggeration to say that Mauss performs a kind of self-analysis before his audience, focusing on the constitution of certain crucial elements of his "habitus" that he shares with other Frenchmen of his generation: "Here let us look for a moment at ourselves. Everything in us all is under command. I am a lecturer for you; you can tell it from my sitting posture and my voice, and you are listening to me seated and in silence. We have a set of permissible or impermissible, natural or unnatural attitudes. Thus we should attribute different values to the act of staring fixedly: a symbol of politeness in the army, and of rudeness in everyday life."[18]

In that sense, Mauss tends to foreground the sociological and psychological processes by which "tradition" leads to the effective transmission of such techniques and the attribution of cultural or social values to them. These processes of imitation and cultivation by which both the infant and the adult learn to execute certain series of movements are crucially mediated by vision: "What takes places is a prestigious imitation. The child, the adult imitates actions which have succeeded and which he has seen successfully performed by people in whom he has confidence and who have authority over him. The action is imposed from without, from above, even if it is an exclusively biological action, involving his body. The individual borrows the series of movements which constitute it [la série des mouvements dont il est composé] from the action executed in front of him or with him by others."[19] While the serial conception of movement is clearly a reference to the analytical breakdown of all sorts of bodily acts

undertaken in the wake of Marey's locomotion physiology by physical anthropologists such as Regnault,[20] Mauss made no reference to their work.
Instead he illustrated his notion of prestige by way of an example that has
become famous, namely, the influence exerted by Hollywood films over
the adoption of certain forms of gait. Lying in a hospital bed in New York
he observes the nurses' peculiar form of walking, which after his return
to France he would also find in the streets of Paris: "American walking
fashions had begun to arrive here, thanks to the cinema."[21] Not without a
certain irony, then, we see that prestige is accorded in this case not to science and its visual analytical devices—the cinematic archive of which the
physical anthropologists dreamt that would encompass the various "languages of gesture" and gait in all its positions—but rather to the synthetic
and commercial form of cinema and its wide-ranging cultural influence.

In considering more closely Mauss's influential teachings on how
to do ethnographic fieldwork, however, one notes a tension between the
imperative to "record it all" and an emphasis on the skills of the human
observer. Thus, while every available recording technique—photography,
phonography, film—is recommended for fixing gestures and movements,
observation by the ethnographer's own senses remains the necessary and
indispensable act of the operation.[22] This tension may explain why the approach adopted by Mauss himself (who did not engage in any fieldwork)
remained closely tied to a certain practice of self-analysis, albeit covering
different material and forms than the psychoanalytical approach advanced
by Freud. It is tempting to recognize here a complementarity between a
new approach toward techniques of the body and another one toward techniques of the mind (notably the technique of "free association"), whose
connections are yet to be brought fully to light. The process of symbolization, identified by Freud on the level of the formation of neurotic symptoms and later by Mauss on the level of socialization, is only the most
visible link between them. Even if the human sciences would not entirely
give up the dream of a complete archive of human gestures and movements, their observational methods remain haunted by the problem of
self-reflexivity. Reflecting on the increasing turn of biomechanical locomotion research during the twentieth century toward the development of
three-dimensional simulation, one might ask to what extent the human
sciences are ineluctably attached to the observation of the real.

ACKNOWLEDGMENTS

The initial project that led to this book started in 2001 at the Max Planck Institute for the History of Science (Berlin) and was completed there, after a detour via Cambridge and Paris, in 2013. During my time at the Department of History and Philosophy of Science at the University of Cambridge (2005–2007), my work was supported by a fellowship of the Sir Henry Wellcome Trust.

The library service of the Max Planck Institute for the History of Science (Berlin) provided invaluable assistance in the original research leading to this book. I would like to thank in particular Ellen Garske and Ruth Kessentini, and the library's former director, Urs Schöpflin.

The first readers of the German manuscript were Lorraine Daston, Doris Kaufmann, and Hans-Jörg Rheinberger, to whom I am very much indebted for their comments.

This new English expanded and revised version was much improved by the criticism and suggestion of an anonymous reader for the University of Chicago Press. I am very grateful to the translators who responded to all my questions with patience and skill. A few sections had to be completely rewritten, such as the introduction, several sections in chapter 4, and the conclusion. Some of the research that has contributed to this book has appeared earlier in "The Physiological Circus: On Knowing, Representing, and Training Horses in Motion in Nineteenth-Century France," *Representations* 111 (Summer 2010): 88–120.

In particular, I would like to thank my editor, Karen Darling, for her patience and for handling the final manuscript with such enthusiasm and efficiency.

Over the years, I have been able to discuss the project with many colleagues and friends (too many to be named here) coming from very different disciplinary backgrounds and intellectual traditions. It is certainly not surprising that each of them had contrasting ideas about the book when it was still in progress. The most important inspirations came, as in the case of my previous book, from the late Lydia Marinelli, who made me read Balzac's *Théorie de la démarche*, as well as from the late John Forrester and Simon Schaffer, who both seemed never tired of exploring in conversation the multifarious possible paths my project could have taken. Whether the path I have ultimately followed in this book can lead to new insights will be decided by its readers.

The translation of this work was funded by Geisteswissenschaften International–Translation Funding for Humanities and Social Sciences from Germany, a joint initiative of the Fritz Thyssen Foundation, the German Federal Foreign Office, the collecting society VG WORT, and the German Publishers and Booksellers Association, and by the Centre Alexandre Koyré (Centre national de la recherche scientifique [CNRS]).

NOTES

INTRODUCTION

1. Both of these widely used expressions too strongly suggest a unilateral process of a society driven by the forces of economy, science, and technology. For some of the recent literature, see Peter Borscheid, *Das Tempo-Virus: Eine Kulturgeschichte der Beschleunigung* (Frankfurt: Campus 2004); François Caron, *La dynamique de l'innovation: Changement technique et changement social (XVIe–XXe siècle)* (Paris: Gallimard, 2010), especially 370–73; Jürgen Osterhammel, *The Transformation of the World: A Global History of the Nineteenth Century* (Princeton, NJ: Princeton University Press, 2014 [2009]), 67–76.

2. Rebecca Solnit, *Wanderlust: A History of Walking* (New York: Penguin Books, 2001), 10.

3. Frederic Gros, *A Philosophy of Walking* (London: Verso, 2015 [2009]), 6–7.

4. "And the true Goddess was revealed with her step" (my trans. of Virgil, Aeneid 1.405). The entire passage reads as follows: "She spake, and as she turned away, her roseate neck flashed bright. From her head her ambrosial tresses breathed celestial fragrance; down to her feet fell her raiment, and in her step she was revealed, a very goddess." Virgil, *Virgil in Two Volumes*, trans. H. Rushton Fairclough (Cambridge, MA: Harvard University Press, 1960), 1:269.

5. See Jan Bremmer, "Walking, Standing, and Sitting in Ancient Greek Culture," in *A Cultural History of Gesture from Antiquity to the Present Day*, ed. Jan Bremmer and Herman Roodenburg (Ithaca, NY: Cornell University Press, 1991), 15–35. For the gestural cultures of the middle ages, see Jean-Claude Schmitt, "The Rationale of Gestures in the West: Third to Thirteenth Centuries," in Bremmer and Roodenburg, *Cultural History of Gesture*, 59–70, and *La raison des gestes dans l'Occident médiéval* (Paris: Gallimard, 1990).

6. Bernd Jürgen Warneken, "Biegsame Hofkunst und aufrechter Gang: Körpersprache und bürgerliche Emanzipation um 1800," in *Der aufrechte Gang: Zur Symbolik einer Körperhaltung* (Tübingen: Ludwig Uhland Institut, 1990), 11–23; Gudrun M. König, *Eine Kulturgeschichte des Spaziergangs: Spuren einer bürgerlichen Praktik*

1780–1850 (Cologne: Böhlau, 1996); Laurent Turcot, *Le promeneur à Paris au XVIIIe siècle* (Paris: Gallimard, 2007).

7. See especially Foucault's influential account in *Discipline and Punish: The Birth of the Prison*, transl. A. Sheridan (New York: Vintage Books, 1995 [1977]). For more recent histories of the body, some of which are indebted to Foucault, see Alain Corbin, Jean-Jacques Courtine, and Georges Vigarello, eds., *Histoire du corps*, 3 vols. (Paris: Le Seuil, 2011); Linda Kalof and William Bynum, eds., *A Cultural History of the Human Body*, 6 vols. (London: Blackwell, 2010); Philipp Sarasin and Jakob Tanner, eds., *Physiologie und industrielle Gesellschaft: Studien zur Verwissenschaftlichung des Körpers im 19. und 20. Jahrhundert* (Frankfurt: Suhrkamp, 1998); Georges Vigarello, *Le corps redressé: Histoire d'un pouvoir pédagogique* (Paris: Armand Colin, 2001 [1978]).

8. Honoré de Balzac, *Ferragus*, in *Oeuvres complètes*, vol. 5 (Paris: Gallimard, 1977), 798; and *Ferragus, Chief of the Dévorants*, transl. Katharine Prescott Wormeley (Gloucester: Project Gutenberg, 2004), 12.

9. Such as sleeping and dreaming, processes that were investigated throughout the twentieth century by a variety of disciplines. For interrelations between studies of locomotion and dreams, see Andreas Mayer, "Des rêves et des jambes: le problème du corps rêvant (Mourly Vold, Freud, Michaux)," *Romantisme* 178, no. 4 (2017): 75–85.

10. Marcel Mauss, "Les techniques du corps," in: *Sociologie et anthropologie*, ed. Claude Lévi-Strauss (Paris: Presses universitaires de France, 1950 [1935]), 363–72. Published in English as "Techniques of the Body," *Economy and Society* 2, no. 1 (1973): 70–88. For a more recent edition, see Marcel Mauss, *Techniques, Technology, and Civilization*, ed. Nathan Schlanger (New York: Berghahn Books, 2006), with an insightful introduction by the editor (1–29).

11. For a first formulation of the problem and methodology, see Andreas Mayer, "Faire marcher les hommes et les images: Les artifices du corps en movement," *Terrain* 46 (March 2006): 33–48.

12. Giovanni Alfonso Borelli, *De motu animalium* (Rome: Angeli Bernabò, 1680–1681); *De motu animalium. Editio nova, a plurimis mendis repurgata* (The Hague: Petrum Gosse, 1743). See the rather problematic English translation by Paul Maquetin, *On the Movement of Animals* (New York: Springer, 1989).

13. For the emphasis on the individual, see Carlo Ginzburg, "Clues: Roots of an Evidential Paradigm," in *Clues, Myths, and the Historical Method* (Baltimore: Johns Hopkins University Press, 1989), 96–125. In later contributions, Ginzburg has also addressed the problem of the production of types and of series. See especially Carlo Ginzburg, "Family Resemblances and Family Trees: Two Cognitive Metaphors," *Critical Inquiry* 30, no. 3 (Spring 2004): 537–56, and "Réflexions sur une hypothèse," in *Mythes, emblèmes, traces: Morphologie et histoire* (Lagrasse: Verdier 2010), 351–64.

CHAPTER 1

1. Jean-Jacques Rousseau, *Julie, or The New Heloise: Letters of Two Lovers Who Live in a Small Town at the Foot of the Alps*, trans. Philip Stewart and Jean Vaché (Dartmouth, NH: Dartmouth College Press, 1997), 207.

2. See Bernd Jürgen Warneken, "Bürgerliche Gehkultur in der Epoche der Französischen Revolution," *Zeitschrift für Volkskunde* 85 (1989): 177–87.

3. See Laurent Turcot, *Le promeneur à Paris au XVIIIe siècle* (Paris: Gallimard, 2007), chap. 1.

4. Many studies discuss the travel literature of the eighteenth century, but most focus either on its status as source material for social and cultural history or on genre-specific questions. See Hans Joachim Piechotta, ed., *Reise und Utopie: Zur Literatur der Spätaufklärung* (Frankfurt: Suhrkamp, 1976); William E. Stewart, *Die Reisebeschreibung und ihre Theorie im Deutschland des 18. Jahrhunderts* (Bonn: Bouvier, 1978); Wolfgang Griep, "Reiseliteratur im späten 18. Jahrhundert," in *Deutsche Aufklärung bis zur Französischen Revolution: 1680–1789*, ed. Rolf Grimminger (Munich: Hanser, 1980), 739–64; and Wolfgang Griep and Hans-Wolf Jäger, eds., *Reise und soziale Realität am Ende des 18. Jahrhunderts* (Heidelberg: Winter, 1983).

5. Jean-Jacques Rousseau, *The Confessions and Correspondence, Including the Letters to Malesherbes*, trans. Christopher Kelly and ed. Christopher Kelly, Roger D. Masters, and Peter G. Stillman (Hanover: University Press of New England, 1995), 136.

6. Most powerfully, perhaps, in the peripatetic poems of Wordsworth. See Anne D. Wallace, *Walking, Literature, and English Culture: The Origins and Uses of Peripatetic in the Nineteenth Century* (Oxford: Clarendon, 1993). Another prominent but little-known representative of this school in late eighteenth-century England was John Thelwall; see his *The Peripatetic*, ed. Judith Thompson (Detroit: Wayne State University Press, 2001 [1793]).

7. On the *Discours* as a core text in the history of anthropology, see Claude Lévi-Strauss, "Jean-Jacques Rousseau, Founder of the Sciences of Man," in *Structural Anthropology*, vol. 2, trans. Monique Layton (Chicago: University of Chicago Press, 1976), 33–43. See also Michèle Duchet, *Anthropologie et histoire au siècle des Lumières* (Paris: Albin Michel, 1995 [1971]), 322–76.

8. Jean-Jacques Rousseau, *Discourse on the Origin of Inequality*, trans. Donald A. Cress (Indianapolis, IN: Hackett, 1992), 19.

9. "Even the external figure of the human species, declares them to be sovereigns of the earth. The body of man is erect; his attitude is that of command; and his countenance, which is turned towards the heavens, is impressed with the signatures of superior dignity." Man's erect "majestic deportment" and the "firmness of his movements" herald the "superiority of his rank." Georges Louis Leclerc comte de Buffon, *Natural History, General and Particular, by the Count de Buffon*, trans. William Smellie, vol. 2 (London: W. Strahan and T. Cadell, 1785), 437. For the relationship between Rousseau and Buffon, see Otis Fellows, "Buffon and Rousseau: Aspects of a Relationship," *Publications of the Modern Language Association of America* 75, no. 3 (1960): 184–96; and Jean Starobinski, "Rousseau and Buffon," in *Jean-Jacques Rousseau: Transparency and Obstruction*, trans. Arthur Goldhammer (Chicago: University of Chicago Press, 1988), 323–32.

10. See Georges Pire, "Jean-Jacques Rousseau et les relations de voyages," *Revue d'histoire littéraire de la France* 3 (1956): 355–78.

11. Rousseau, *Discourse*, 20.

12. Ibid., 24.

13. Jean-Jacques Rousseau, *Émile, or Education*, trans. Barbara Foxley (London: J. M. Dent and Sons, 1921; New York: E.P. Dutton, 1921), 139. This passage should be read not as a general criticism of the natural sciences but in relation to the important role of the scientist's body in scientific self-experiments. On this topic, see especially Simon Schaffer, "Self Evidence," *Critical Inquiry* 18, no. 2 (1992): 327–62.

14. Rousseau, *Émile*, 90.

15. Ibid., 374–75.

16. Cuvier's manuscript (his first known text) of the *Voyage dans les Alpes souabes* has been edited in Philippe Taquet, *"Les premiers pas d'un naturaliste sur les sentiers du Wurtemberg: Récit inédit d'un jeune étudiant nommé Georges Cuvier." Geodiversilas 20 (2): 285–318*. See also chapters 15 and 16 in Philippe Taquet, *Georges Cuvier: Naissance d'un genie* (Paris: Odile Jacob, 2006); and Alphonse Favre, *H.-B. de Saussure et les Alpes* (Lausanne: Bridel, 1870). In Favre's hagiography, Saussure appears as the founder of an entirely new method of observation developed through the act of walking.

17. Heinrich August Ottokar Reichard, *Der Passagier auf der Reise in Deutschland und einigen angränzenden Ländern, vorzüglich in Hinsicht auf seine Belehrung, Bequemlichkeit und Sicherheit: Ein Reisehandbuch für Jedermann* (Weimar: Gädicke 1801), 85.

18. See the definition of "Maschine" in Krünitz's *Ökonomisch-technologische Encyklopädie*: "in its true sense any artificially assembled object without life or its own movement. More narrowly, such an assembly, but with a purpose; a *tool*. For example, the *tobacco machine* for smoking tobacco, where the smoke, before it reaches the mouth, is guided through clear water. The *tea machine, etc*. In another narrow sense, the *machine* is an artificially assembled object, provided with some movement, but unable to move on its own. The world-edifice, the clock, etc. are such machines. In the narrowest sense it is an assembled tool that helps create or facilitate a movement, in contrast to a mere *tool* or *instrument*, which may also be simple." Krünitz, *Ökonomisch-technologische Encyklopädie* (Berlin: J. Pauli, 1802), 85:160, http://www .kruenitz1.uni-trier.de.

19. This may be why its history has only been investigated sporadically. For a concise sociohistorical sketch, see Wolfgang Kaschuba, "Die Fußreise: Von der Arbeitswanderung zur bürgerlichen Bildungsbewegung," in *Reisekultur: Von der Pilgerfahrt zum modernen Tourismus*, ed. Hermann Bausinger, Klaus Beyrer and Gottfried Korff (Munich: C. H. Beck, 1991), 165–73.

20. See Klaus Beyrer, *Die Postkutschenreise* (Tübingen: Tübinger Vereinigung für Volkskunde, 1985); Thomas Brune, "Von Nützlichkeit und Pünktlichkeit der Ordinari-Post," in Bausinger, Beyrer, and Korff, *Reisekultur*, 123–30; for France, as compared to Germany and the British Isles, see Christophe Studeny, *L'invention de la vitesse: France, XVIIIe–XXe siècle* (Paris: Gallimard, 1995), 172–93.

21. See, for example, Reichard, *Der Passagier auf der Reise in Deutschland*, 117. See also Balzac's later dramatic report on the different coaches he had to take on his journey from Paris to Russia: Honoré de Balzac, *Abglanz meines Begehrens: Bericht einer Reise nach Russland 1847* (Berlin: Friedenauer Presse, 2018).

22. [Johann Kaspar Riesbeck], *Briefe eines reisenden Franzosen über Deutschland an seinen Bruder zu Paris*, 2nd ed. (Zürich, 1784), 1:14.

23. Friedrich Nicolai, *Beschreibung einer Reise durch Deutschland und die Schweiz im Jahre 1781. Nebst Bemerkungen über Gelehrsamkeit, Industrie, Religion und Sitten*, 3rd ed. (Berlin, 1788), 1:4–18. On Nicolai's account of his travels and its paradoxes, see Hans Joachim Piechotta, "Erkenntnistheoretische Voraussetzungen der Beschreibung: Friedrich Nicolais Reise durch Deutschland und die Schweiz im Jahre 1781," in Piechotta, *Reise und Utopie*, 98–150.

24. "Beschreibung des Catelschen an einen Wagen angebrachten Wegmessers; nebst gesammelten Nachrichten von einigen ältern Werkzeugen dieser Art." Nicolai, *Beschreibung*, app. 1.

25. See Johann Beckmann, *Beyträge zur Geschichte der Erfindungen* (Leipzig: Paul Gotthelf Kummer, 1782), 16–27; Jacob Leupold, "Von den Wagen-Instrumenten," in *Theatri Machinarum Supplementum: Das ist: Zusatz zum Schauplatz der Machinen und Instrumenten* (Leipzig: Breitkopf, 1739), 22–28; Krünitz, *Ökonomisch-technologische Encyklopädie*, s.v. "Wegmesser," 235:496–97; Heinrich August Ottokar Reichard, *Handbuch für Reisende aus allen Ständen* (Leipzig: Weygandtschen Buchhandlung, 1784), 327–31.

26. Nicolai: *Beschreibung*, 16.

27. Rousseau, *Émile*, 374.

28. Ibid.

29. Ibid., 373.

30. [Jonas Ludwig von Heß], *Durchflüge durch Deutschland, die Niederlande und Frankreich* (Hamburg: Bachmann und Gundermann, 1793), 1:5–16. See also Joist Grolle, "Republikanische Wanderungen: Die Fußreisen des Jonas Ludwig von Heß aus Hamburg durch die 'Freien deutschen Reichsstädte' 1789–1800," *Zeitschrift des Vereins für Hamburgische Geschichte* 83, no. 1 (1997), 299–321.

31. Karl Philipp Moritz: "Reisen eines Deutschen in England im Jahr 1782," in *Popularphilosophie, Reisen, Ästhetishe Theorie*, vol. 2 of *Werke in zwei Bänden*, ed. Heide Hollmer and Albert Meier (Frankfurt: Deutscher Klassiker, 1997), 309. See also page 307: "It was a beautiful day, the most gorgeous vistas on either side, upon which the eye would gladly have rested longer, if our coach had not so jealously rolled past it all."

32. Ibid., 312.

33. Ibid., 321.

34. Ibid., 340.

35. Reichard, *Der Passagier auf der Reise in Deutschland*, 86.

36. Heinrich Ludwig Christian Böttger, "Vorschlag einer Uniform für Reisende zu Fuße," *Journal des Luxus und der Moden* 15 (May 1800): 217.

37. Letter from Vieth to his father, Dessau, October 17, 1790, in Gustav Krüger, ed., *Zur Erinnerung an Gerhard Anton Ulrich Vieth, weiland Schulrat und Direktor der Herzogl. Hauptschule zu Dessau. 1786–1836. Aus seinem Nachlass* (Dessau: Paul Baumann, 1885), 40. Krünitz, too, in his article on the "Cabriolet," emphasizes how uncomfortable these vehicles were (Krünitz, *Ökonomisch-technologische Encyklopädie*, 7:501): "Because cabriolets hang from neither springs nor straps, and their movement has thus been found unyielding, the saddlers have tried to ameliorate this inconvenience by constructing the seat as a two-layered banquette, where the upper part, on which the pillow lies, rests on cylinders of spiral steel springs."

38. Jacques-Louis de Latocnaye, *Meine Fußreise durch Schweden und Norwegen: Ein Seitenstück zu der Reise des Verfassers durch die drey brittischen Königreiche; Mit Anmerkungen und Zusätzen eines Deutschen*, trans. Eduard Henke (Leipzig: Hartknoch, 1802), 2:41.

39. Johann Gottfried Seume, *Spaziergang nach Syrakus im Jahre 1802*, ed. Albert Meier, 3rd ed. (Munich: Deutscher Taschenbuch, 1994), 68.

40. Ibid., 68–69.

41. Otto Spazier, review of Seume, *Spaziergang nach Syracus. Zeitung für die elegante Welt*, June 4, 1803.

42. Johann Gottfried Seume, "Mein Sommer 1805," in *Werke in zwei Bänden*, ed. Jörg Drews (Frankfurt: Deutscher Klassiker Verlag, 1993), 1:543.

43. See Philippe Ariès, *Centuries of Childhood: A Social History of Family Life*, trans. Robert Baldick (New York: Alfred A. Knopf, 1962), 50–61.

44. Rousseau, *Émile*, 42n1.

45. Krünitz, *Ökonomisch-technologische Encyklopädie*, s.v. "Gängeln," 15:626.

46. Ibid., 626–7.

47. Christian Gotthilf Salzmann, "Wie gut es sey, seine Kinder das Gehen selbst lernen zu lassen," in *Nachrichten aus Schnepfenthal für Eltern und Erzieher* (Leipzig: Siegfried Lebrecht Crusius, 1786), 1:168–73. That these methods of physical pedagogy were also projected into the moral realm can be seen in admonitions to abolish completely all "moral leading strings," commandments, and prohibitions as a means of fostering goodness in a person. See Christian Gotthilf Salzmann, *Ameisenbüchlein, oder Anweisung zu einer vernünftigen Erziehung der Erzieher* (Reutlingen: Mäcken, 1808), 81.

48. Johann Christoph Friedrich Gutsmuths, *Gymnastik für die Jugend*, 2nd. ed. (Schnepfenthal: Buchhandlung der Erziehungsanstalt, 1804), 187.

49. Ibid.

50. Ibid.

51. Peter Burke, *The Fortunes of the Courtier: The European Reception of Castiglione's Cortegiano* (London: Polity, 1995).

52. For an older overview, see Hajo Bernett, *Die pädagogische Neugestaltung der bürgerlichen Leibesübungen durch die Philanthropen*, 3rd. ed. (Schorndorf bei Stuttgart: Hoffmann, 1971). For more recent historical studies on ideals of posture since the early modern period, see Herman Roodenburg, *The Eloquence of the Body: Perspectives on Gesture in the Dutch Republic* (Zwolle: Waanders, 2004); and Kirsten O. Frieling, "Haltung bewahren: Der Körper im Spiegel frühneuzeitlicher Schriften über Umgangsformen," in *Bewegtes Leben: Körpertechniken in der frühen Neuzeit*, ed. Rebekka von Mallinckrodt (Wolfenbüttel: Herzog August Bibliothek, 2008), 39–59.

53. Gutsmuths, *Gymnastik für die Jugend*, 186.

54. On the history of court dance, see Rudolf Braun and David Gugerli, *Macht des Tanzes, Tanz der Mächtigen: Hoffeste und Herrschaftszeremoniell 1550–1914* (Munich: C. H. Beck, 1993); and Stephanie Schroedter, Marie-Thérèse Mourey and Gilles Bennett, eds., *Barocktanz im Zeichen französisch-deutschen Kulturtransfers: Quellen zur Tanzkultur um 1700* (Hildesheim: Olms, 2008).

55. See "Von der Tanzkunst," "Von der Fechtkunst," and "Von der Reitkunst," in Gerhard Ulrich Anton Vieth, *Versuch einer Encyclopädie der Leibesübungen*, ed. Friedrich Fetz (Frankfurt: Limpert, 1970 [1795]), 2:150–86, 203–29, and Gutsmuths, *Gymnastik für die Jugend*, 283–300, 326–55.

56. Gutsmuths, *Gymnastik für die Jugend*, 185–86.

57. Vieth, *Versuch*, 2:152–3. Here Vieth follows Jean Georges Noverre, *Lettres sur la danse et sur les ballets* (Lyon: Aimé Delaroche, 1760), 315–61 (letter 12), who criticized the hip turner as a "badly thought-out and badly executed machine that, so far from working effectively, maims those who make use of it, by imprinting a defect upon the waist that is much more disagreeable than the one it wants to destroy" (321).

58. Vieth, *Versuch*, 2:153.

59. Ibid., 29.

60. Ibid., 167n98.

61. Ibid., 79.

62. Friedrich Wilhelm Joseph von Schelling, "Von der Weltseele" (1798), in *Friedrich Wilhelm Joseph von Schellings sämmtliche Werke: Erste Abtheilung* (Stuttgart and Augsburg: Cottascher, 1857), 2:540.

63. Giovanni Alfonso Borelli, *De motu animalium* (Rome: Angelo Bernabò, 1680–1681). Only part of the work was translated into German, by Silvester Heinrich Schmidt, as *Von der wundersamen Macht der Muskuln* (1706).

64. Johann Gottlob Krüger, *Naturlehre. Zweyter Theil, welcher die Physiologie, oder Lehre von dem Leben und der Gesundheit der Menschen in sich fasset* (Halle: Hermann Hemmerde, 1748), 807. Krüger's critical insight can be expanded to include the continued reception of Borelli's work, in which his own physiology played a significant role. At least in Germany, until the early nineteenth century, his brief chapter on Borelli, "Von der Würkung der Muskeln," was read much more frequently than the Latin original. On Krüger and medical anthropology more generally, see Carsten Zelle, ed., *"Vernünftige Ärzte": Hallesche Psychomediziner und die Anfänge der Anthropologie in der deutschsprachigen Frühaufklärung* (Tübingen: Niemeyer, 2001).

65. Krüger, *Naturlehre*, 800–801. See also Vieth, *Versuch*, 2:31.

66. [Friedrich Christoph von Saldern], *Taktische Grundsätze und Anweisung zu militairischen Evolutionen: Von der Hand eines berühmten Generals* (Dresden: Walthersche Hofbuchhandlung, 1786 [1781]), 2.

67. Friedrich Christoph von Saldern, *Taktische Grundsätze und Anweisung zu militairischen Evolutionen*, ed. Heinrich Johannes Krebs (Copenhagen: Johann Heinrich Schubothe, 1796), 4, note by Krebs.

68. See Max Jähns, *Geschichte der Kriegswissenschaften vornehmlich in Deutschland* (Munich and Leipzig: R. Oldenbourg, 1891), 3:1813–22. In newer studies of military history, this process is also discussed in the context of a "military Enlightenment." See also Daniel Horath, "Spätbarocke Kriegspraxis und aufgeklärte Kriegswissenschaften: Neue Forschungen und Perspektiven zu Krieg und Militär im 'Zeitalter der Aufklärung,'" *Aufklärung* 12 (2000): 5–47; Daniel Horath, "Die Beherrschung des Krieges in der Ordnung des Wissens: Zur Konstruktion und Systematik der militairischen Wissenschaften im Zeichen der Aufklärung," in *Wissenssicherung, Wissensordnung und*

Wissensverarbeitung, ed. Theo Stammen and Wolfgang Weber (Berlin: Akademie, 2004), 371–86; and Jürgen Lüh, *Kriegskunst in Europa, 1650–1800* (Cologne: Böhlau, 2004).

69. In addition to Saldern's *Taktische Grundsätze* (1796), see Jacques-Antoine-Hippolyte de Guibert, *Essai général de tactique, précédé d'un discours sur l'état actuel de la politique et de la science militaire en Europe; avec le plan d'un ouvrage intitulé: La France politique et militaire,* 2 vols. (Liège: 1770); Jacob Friedrich von Rösch, *Mathematische Säze aus der Tactik* (Stuttgart: Cotta, 1778); Heinrich Johannes Krebs, *Anfangsgründe der eigentlichen Kriegswissenschaft: Aus den besten militärischen Schriften zusammengetragen* (Leipzig: Korte, 1784); and Franz von Miller, *Reine Taktik der Infanterie, Cavallerie und Artillerie,* 2 vols. (Stuttgart: Buchdruckerei der hohen Karlsschule, 1787–1788).

70. Miller, *Reine Taktik,* vol. 2, 4.

71. See Harald Kleinschmidt, *Tyrocinium militare: Militärische Körperhaltungen und -bewegungen im Wandel zwischen dem 14. und dem 18. Jahrhundert* (Stuttgart: Autorenverlag, 1989), chap. 5.

72. *Reglement vor die königl. preußische Infanterie* (1743), quoted in Lüh, *Kriegskunst,* 194.

73. Guibert, *Essai,* 1:53. An English translation appeared under the title *General Essay on Tactics,* 2 vols. (London: J. Millan, 1781). The excerpts presented here have been newly translated.

74. Ibid., 50.

75. Ibid.

76. Ibid., 37.

77. Ibid.

78. Ibid., 38.

79. Ibid., 39.

80. Ibid.

81. Ibid., 51.

82. Ibid., 51–52.

83. Ibid., 54–55.

84. Ibid., 60.

85. For a comparative discussion of German drill books, see Kleinschmidt, *Tyrocinium militare,* 253–7.

86. Guibert, *Essai,* 1:55.

87. [Saldern], *Taktische Grundsätze* (1786), 2.

88. Ibid.

89. Following Foucault's work on power, the construction of a new physical ideal in the second half of the eighteenth century has often been seen as the result of an overarching social process of disciplining, with the military drill as its primary model. According to Foucault, disciplinary power in the form of regular exercises penetrates the body and thus produces a "new object," the "natural body," which is a "bearer of forces and the seat of duration." Michel Foucault, *Discipline and Punish: The Birth of the Prison,* trans. Alan Sheridan (New York: Vintage Books, 1995 [1977]), 155. Foucault not only derives his analysis chiefly from Guibert but also declares Guibert's concept of "tactics" to be the analytical category that best illustrates the production of the

disciplined body (167–68). However, to equate epistemic mediation processes with the workings of a diffuse disciplinary force conceals the problem of how human movement can become an object of study at all.

90. Miller, *Reine Taktik*, 2:19.

91. Ibid., 24. Accordingly, various watches were required for the various march steps: "One certainly cannot achieve full travel step using the same watch that is used for the other steps" (29).

92. Ibid., 25. For some initial forays into the functions of military music, a still largely unresearched field, see Jürgen Elvert and Michael Salewski, eds., *Militär, Musik und Krieg—Kolloquium anlässlich des 70. Geburtstages von Michael Salewski"* Historische Mitteilungen der Ranke-Gesellschaft 22 (Stuttgart: Franz Steiner, 2009).

93. Miller, *Reine Taktik*, 2:16–7.

94. "Had he mentioned some famous men, wherever they were of the same opinion, many a reader would have paid more attention to the subject matter." Anon., "Kriegswissenschaften. Stuttgart, in der Druckerey der hohen Karlsschule: Reine Taktik der Infanterie, Cavallerie und Artillerie, in zwey Theilen verfasst von Franz Miller," *Allgemeine Literatur-Zeitung*, March 8, 1789, 578.

95. Anon., "Antikritik gegen die in der Allg. Deutsch. Bibl. befindliche Recension, über die Reine Taktik des Rittm. und Flgladj. v. Miller," *Intelligenzblatt der Allgemeinen Literatur-Zeitung*, May 19, 1790, col. 506–12.

96. Saldern, *Taktische Grundsätze* (1796), 4; note by Krebs.

97. See, for example, Michael Sikora, "'Ueber die Veredlung des Soldaten': Positionsbestimmungen zwischen Militär und Aufklärung," in *Die Bestimmung des Menschen*, ed. Norbert Hinske (Hamburg: Meiner, 1999), 25–50.

98. Johann Caspar Lavater, *Physiognomische Fragmente zur Beförderung der Menschenkenntnis und Menschenliebe: Vierter Versuch* (Leipzig: Weidmanns Erben und Reich, 1778), 419.

99. For the reception of Lavater's physiognomy, see especially Ellis Shookman, ed., *The Faces of Physiognomy: Interdisciplinary Approaches to Johann Caspar Lavater* (Columbia, SC: Camden House, 1993); and Melissa Percival and Graeme Tytler, eds., *Physiognomy in Profile: Lavater's Impact on European Culture* (Newark: University of Delaware Press, 2005).

100. Lavater, *Physiognomische Fragmente*, 418.

101. "I thus believe (and my everyday observations confirm this abundantly) that a person's gait is of the utmost physiognomic importance, and that a good physiognomist could be capable of safely indicating, or of expressing by imitation, the gait of a proud, meek, spirited, anxious, honest, deceitful, ignorant, prudent, etc. person." Lavater, *Physiognomische Fragmente*, 418–19. This passage also appeared in Lavater's first *Versuch: Von der Physiognomik* (Leipzig: Weidemanns Erben und Reich, 1772), 183.

102. This aesthetics of line calls to mind William Hogarth's "line of beauty"; *The Analysis of Beauty* (London: J. Reeves, 1753).

103. Johann Caspar Lavater, *Vermischte physiognomische Regeln: Ein Manuscript für Freunde* (Zurich: Orell, Füssli, 1802 [1789]), 42–3.

104. See August Ohage, "'Raserei für Physiognomik in Niedersachsen': Lavater, Zimmermann, Lichtenberg und die Physiognomik," in *Georg Christoph Lichtenberg*

1742–1799: Wagnis der Aufklärung (Vienna: Hanser, 1992), 175–84. For Chodowiecki's role, see Thomas Kirchner, "Chodowiecki, Lavater und die Physiognomiedebatte in Berlin," in *Daniel Chodowiecki (1726–1801): Kupferstecher, Illustrator, Kaufmann,* ed. Ernst Hinrichs and Klaus Zernack (Tübingen: Niemeyer, 1997), 101–42.

105. Letter from Lichtenberg to Johann Andreas Schernhagen, October 17, 1775, in *Schriften und Briefe,* ed. Wolfgang Promies (Munich: Hanser, 1972), 4:252.

106. Georg Christoph Lichtenberg, "Natürliche und affektierte Handlungen des Lebens: Erste Folge," in *Der Fortgang der Tugend und des Lasters: Daniel Chodowieckis Monatskupfer zum Göttinger Taschenkalender mit Erklärungen Georg Christoph Lichtenbergs 1778–1783* (Berlin: Der Morgen, 1977), 29.

107. The dual meaning of *Spaziergang* or *Promenade* as "a space for recovery" and a "walk which one makes, or undertakes, only for entertainment, for recreation" (as recorded in Krünitz's *Oeconomisch-technologische Encyklopädie,* 156:587), gained currency in Germany only after 1800. See Gudrun M. König, *Eine Kulturgeschichte des Spaziergangs: Spuren einer bürgerlichen Praktik 1780–1850* (Cologne: Böhlau, 1996), 24–5.

108. The innumerable individual variations of the gait encountered in everyday life could, for the attentive observer, be distinguished not only by sight but also by sound: "Every human has his distinctive walk, by which one can recognize him with the ear alone; proof not only of the sensitivity of the ear but also of the peculiar diversity that arises in the advancement and placement of just two feet." Vieth, *Versuch,* 2:79.

109. In addition to a Dutch translation, a French version appeared under the title *Idées sur le geste et l'action théâtrale* (Paris: Jansen, 1795); an Italian translation by Giovanni Rasori as *Lettere intorno alla Mimica* (Milano: Batelli e Fanfani, 1820); and an English adaptation by Henry Siddons, *Practical Illustrations of Rhetorical Gesture and Action* (Sherwood: Neely and Jones, 1822).

110. Johann Jakob Engel, *Ideen zu einer Mimik: Erster Theil* (Berlin: Mylius, 1785), 23–24.

111. Ibid., 7.

112. "The same in gait and spirit" (from Seneca's tragedy, *Hercules furens*). Ibid., 98.

113. Vieth, *Versuch,* 2:80–81.

CHAPTER 2

1. See Roselyne Rey, *Naissance et développement du vitalisme en France de la deuxième moitié du 18e siècle à la fin du Premier Empire* (Oxford: Voltaire Foundation, 2003 [1987]); Elizabeth Williams, *The Physical and the Moral: Anthropology, Physiology, and Philosophical Medicine in France, 1750–1850* (Cambridge: Cambridge University Press, 1994); and Williams, *A Cultural History of Medical Vitalism in Enlightenment Montpellier* (Burlington, VT: Ashgate, 2003). Despite the obvious problem of gender, the translation "science of man" has been adopted here for *Science de l'homme;* see also Williams, *Physical and the Moral.*

2. Even longer, in fact. Georges Canguilhem, in particular, repeatedly mobilized the vitalist legacy in the history and epistemology of the modern life sciences. See Georges

Canguilhem, "Aspects of Vitalism," in *Knowledge of Life*, trans. S. Geroulanos and D. Ginsburg (New York: Fordham University Press, 2008), 59–74.

3. See Sergio Moravia, "Philosophie et médecine en France à la fin du XVIIIe siècle," *Studies in Voltaire and the Eighteenth Century* 39 (1972), 1089–151; Moravia, *La scienza dell'uomo nel Settecento: Con una appendice di testi* (Rome-Bari: Laterza, 1970); and the detailed study by Jean Luc Chappey, *La Société des observateurs de l'homme (1799–1804): Des anthropologues au temps de Bonaparte* (Paris: Société des études robbespierristes, 2002).

4. Especially in the rather selective reception of Michel Foucault's *The Birth of the Clinic: Archaeology of Medical Perception*, trans. Alan Sheridan (London: Routledge, 1989 [1963]), but also in studies not inspired by Foucault, such as Dora Weiner and Michael Sauter, "The City of Paris and the Rise of Clinical Medicine," *Osiris* 18 (2003): 23–42.

5. François Boissier de Sauvages, *Nosologia methodica sistens morborum classes, genera et species, juxta Sydenhami mentem et Botanicorum ordinem* (Amsterdam: Frères De Tournes, 1763); Boissier de Sauvages, *Nosologie méthodique, dans laquelle les maladies sont rangées par classes, suivant le système de Sydenham, & l'ordre des botanists*, 10 vols. (Paris: Hérissant le fils, 1771); Philippe Pinel, *Nosographie philosophique, ou la méthode de l'analyse appliquée à la medicine* (Paris: Richard, Caille and Ravier, 1798).

6. Paul-Joseph Barthez, *Nouvelle mécanique des mouvements de l'homme et des animaux* (Carcassonne: de l'Imprimerie de Pierre Polère, 1798), xiii. The book drew on articles Barthez had published in the *Journal des Sçavans* between 1782 and 1787.

7. A detailed study of the reception of this important work has yet to be undertaken. Several editions of the Latin text appeared between 1710 and 1743, edited by the mathematician Johann Bernoulli (1667–1748), who taught in Basel and had studied with Borelli.

8. [Arnulphe d'Aumont], "Debout," in *Encyclopédie ou Dictionnaire raisonné des sciences, des arts et des métiers*, ed. Denis Diderot and Jean-Baptiste le Rond d'Alembert (Paris: Briasson, 1754), 4:654–57. The two other key references for d'Aumont were Haller and Winslow.

9. See, for example, Eyvind Bastholm, *The History of Muscle Physiology* (Copenhagen: Munksgaard, 1950), 165–74; Lester Snow King, *The Philosophy of Medicine: The 18th Century* (Cambridge, MA: Harvard University Press, 1978), 102–9; on mechanical physiology generally, see Theodore M. Brown, *The Mechanical Philosophy and the "Animal Oeconomy"* (New York: Arno, 1981).

10. Aumont, "Debout," 654.

11. Ibid., 656.

12. Ibid., 657.

13. See Federico di Trocchio, "Barthez et l'Encyclopédie," *Revue d'histoire des Sciences* 34, no. 2 (1981), 123–36. Trocchio considers these years central for Barthez's final turn toward an "empirisme raisonné."

14. The vital principle that Barthez defined in his great work *Nouveaux eléments de la Science de l'homme* (Montpellier: Jean Martel, 1778) is conceived as an antimetaphysical abstraction: "A being whose unity and elements one can recognize and whose

material form of existence one does not know, although its existence shows itself in an uncountable number of facts" (1:41). But as Barthez explained in a later edition of the book, the vital principle was not meant to be a causal explanation for the whole great variety of organic phenomena (*Nouveaux eléments*, 3rd ed. [Paris: Germer Ballière: 1858], 1:2). The identical passage also appears in the second edition, from 1806 (page 2).

15. Barthez, *Nouvelle mécanique*, viii.

16. Ibid., 55. Borelli's theory is explained in part 1, chapter 19, "De gressu bipedum," in *De motu animalium: Editio nova, a plurimis mendis repurgata* (The Hague: Petrum Gosse, 1743)159–67.

17. Barthez, *Nouvelle mécanique*, 53.

18. Ibid., iii.

19. Barthez, "Force des animaux," in *Encyclopédie ou Dictionnaire raisonné des sciences, des arts et des métiers*, ed. Denis Diderot and Jean-Baptiste le Rond d'Alembert (Paris: Briasson, 1754), 7:124; reprinted in Barthez, *Nouveaux éléments*, 2:476–96.

20. Barthez, *Nouveaux éléments*, 77.

21. Ibid., 80. The disease, which had been previously described by Gaubius, was named "Scelotyrbe *festinans*" by Boissier de Sauvages in 1763. In 1817, James Parkinson took up these nosological descriptions in the second chapter of *An Essay of the Shaking Palsy*. He defined them as characteristic for one of the stages of the neurological disorder that now bears his name.

22. See Lorraine Daston, "The Empire of Observation, 1600–1800," in *Histories of Scientific Observation*, ed. Lorraine Daston and Elizabeth Lunbeck (Chicago: University of Chicago Press, 2011), 81–113, especially 86–87. This observation applies particularly to d'Alembert, which, however, did not prevent Barthez from repeatedly citing him as an authority along with Bacon and Newton. See the expanded version of the "Discours préliminaire" in the second (1806) edition of the *Nouveaux Éléments*; also in the third edition (1858, 1–41).

23. [Ménuret de Chambaud], "Observateur," *Encyclopédie ou Dictionnaire raisonné des sciences, des arts et des métiers*, ed. Denis Diderot and Jean-Baptiste le Rond d'Alembert (Paris: Briasson, 1751), 11: 310. See also his article "Observation," in the same volume. On the physicians who contributed to the *Encyclopédie*, see Jacques Roger, *Les sciences de la vie dans la pensée française au XVIIe siècle* (Paris: Albin Michel, 1993 [1963]), 630–39.

24. See Hans Blumenberg, *Die Lesbarkeit der Welt*, 2nd ed. (Frankfurt: Suhrkamp, 1983) and Pierre Hadot, *The Veil of Isis: An Essay on the History of the Idea of Nature*, trans. Michael Chase (Cambridge, MA: Harvard University Press, 2006).

25. Barthez, *Nouvelle mécanique*, xii.

26. On the methods of the antiquarians, see Arnaldo Momigliano, "Ancient History and the Antiquarian," *Journal of the Warburg and Courtauld Institutes* 13, no. 3/4 (1950): 285–315.

27. Barthez, *Nouvelle mécanique*, 20n1.

28. See François Azouvi, *Maine de Biran: La Science de l'homme* (Paris: Vrin, 1995); Martin Staum, "Cabanis and the Science of Man," *Journal of the History of the Behavioral Sciences* 10, no. 1 (1974): 135–43; Williams, *Physical and the Moral*, 213–24.

29. Charles-Louis Dumas, "Observation sur le squelette d'un sauteur, dont les membres abdominaux (extrémités inférieures), étoient composés d'une seule pièce et du pied: suivie de quelques réflexions sur la théorie du saut," in *Recueil périodique de la Société de médecine de Paris* 10 (1800): 30–35, 34ff.; see also Dumas, *Principes de physiologie, ou Introduction à la science expérimentale, philosophique et médicale de l'homme vivant* (Paris: Deterville, 1800), 3:167–68.

30. Michel de Montaigne, *Les essais* (Paris: Gallimard 2007), 114. The English translation is from chapter 22, "Of custom, and that we should not easily change a law received," in *Essays of Michel de Montaigne*, trans. Charles Cotton and ed. William Carew Hazlitt (London: Reeves and Turner, 1877), https://www.gutenberg.org/files/3600/3600-h/3600-h.htm#link2HCH0022.

31. Paul-Joseph Barthez, "Éclaircissemens sur quelques points de la mécanique des mouvemens de l'homme," *Mémoires de la Société médicale d'emulation* 11 (1803): 269.

32. Charles-Louis Dumas, *Principes de physiologie*. 2nd ed. (Paris: Méquignon, 1806), 4:289.

33. Ibid., 290.

34. Barthez's attempt to marry a mechanics of human and animal locomotion with the doctrine of the vital principle was also criticized outside medical circles. A lack of detailed observations, a tendency toward sweeping statements, and an opaque style crowded with abstractions and generalizations—all of these hindered any wide dissemination of Barthez's works on physiology. In 1821, for example, a French biographical dictionary of notable persons remarked, "Barthez is less fortunate when he gets down to details. By nature, his spirit was less suited to tasks associated with the attentive observation of facts." He was granted "an extraordinary talent for generalization," but that was something of an ambiguous compliment: "he pursued physiology not by means of observation but with genius; he divined it, if the expression is permitted." "Barthez," in A. V. Arnault et al., *Nouvelle biographie des contemporains* (Paris: Libraire Historique, 1821), 2:175.

35. Xavier Bichat, *Recherches physiologiques sur la vie et la mort* (Paris: Brosson, Gabon, 1800). For a detailed discussion, see Rey, *Naissance et développement du vitalisme*, 323–72.

36. Matthieu François-Régis Buisson, *De la division la plus naturelle des phénomènes physiologiques considérés chez l'homme, avec un précis historique sur M. F. X. Bichat* (Paris: Brosson, 1802), 110.

37. "In animals, . . . whose faces usually show little expression, the passions are communicated through the gestures of the various parts of the body. . . . Who has not observed a hundred times the particular gestures of each species that establish the relationship between male and female that precedes mating? . . . When they are close to one another, gesture is the unique language with which they paint their mutual condition, and as this gesture is not a matter of convention, each species has its own. It is a mode of communication governed by instinct in which the cerebral functions play no part." Xavier Bichat, *Anatomie descriptive*, rev. ed. (Paris: Gabon, 1829), 2:94–95.

38. Buisson, *De la division la plus naturelle des phénomènes physiologiques*, 134.

39. Ibid., 130–31.

40. Ibid., 134.

41. Denis Diderot, *Lettre sur les aveugles à l'usage de ceux qui voient: Lettre sur les sourds et muets à l'usage de ceux qui entendent et qui parlent* (Paris: Flammarion 2000). The Institut national des sourds et muets, headed from 1789 by Abbé Sicard, held public demonstrations that attracted many Parisian tourists. So, too, did the Institut royal des jeunes aveugles, founded by Valentin Haüy in 1784.

42. Thus, future advances in the science of man could come only by fumbling through an uncharted terrain, one that could not be surveyed with the instruments of natural science. See, for example, Charles-Louis Dumas, *Discours sur les progrès futurs de la Science de l'homme* (Montpellier: Tournel, 1804), 27–28: "The science of man tackles a much too complicated matter; it encompasses too many and too diverse facts; it treats too many and too subtle elements to be able to bring to this infinity of possible combinations the unity, evidence, and certainty that are the hallmark of the physical and mathematical sciences."

43. Pierre Joseph Rullier, "Geste," in *Dictionnaire des sciences médicales*, 60 vols. (Paris: Panckoucke, 1812–1822), 18:329–79; "Locomotion," 28:548–78; "Marche," 31:6–23; "Mouvement," 34:438–59; "Motilité34:401–3. The term *motilité* comes from François Chaussier, *Table synoptique des propriétés caractéristiques et des principaux phénomènes de la force vitale*, 2nd ed. (Paris: Théophile Barrois, 1800).

44. Rullier, "Geste," 359.

45. Rullier, "Locomotion," 578.

46. Rullier, "Geste," 375.

47. On the importance of classical rhetoric for early nineteenth-century French psychiatry, see Juan Rigoli, *Lire le délire: Aliénisme, rhétorique et littérature en France au XIXe siècle* (Paris: Fayard, 2001), especially chapter 2.

48. Étienne Esquirol, "Manie," in *Dictionnaire des sciences médicales* (Paris: Panckoucke, 1818), 30:446.

49. Ibid., 449.

50. See "Moreau de la Sarthe," in *Dictionnaire encyclopédique des sciences médicales*, ed. Amédée Dechambre, Léon Lereboullet, and Louis Hahn (Paris: Masson, 1876), 357–58.

51. For the reception of physiognomy in Paris, see Martin Staum, "Physiognomy and Phrenology at the Paris Athénée," *Journal of the History of Ideas* 56, no. 3 (1995): 443–62.

52. Louis-Jacques Moreau de la Sarthe, ed., *L'art de connaître les hommes par la physionomie par Gaspard Lavater*(Paris: 1820), 1:101, 103.

53. Ibid., 137.

54. Ibid., 138.

55. Engel's book had been translated into French as early as 1795, but Moreau appears to have used the German version of the text, published in 1804 in volumes 7 and 8 of Engel's *Schriften*. See Moreau de la Sarthe, ed., *L'art de connaître les hommes*, 3:18.

56. Ibid., 15.

57. Ibid., 17. Elsewhere, he calls imitation an "incalculable force": "Observations sur les signes physionomiques des professions," in ibid., 6:225.

58. "Réflexions sur les caractères physionomiques tirés de la forme de l'écriture par l'un des éditeurs [1806]," in ibid., 3:122.

59. Engel, *Ideen zu einer Mimik* (Berlin: Mylius, 1785), 92–93; Moreau de la Sarthe, *L'art de connaître les hommes*, 3:28, 31.

60. Moreau de la Sarthe, "Observations sur les signes physionomiques des professions," in Moreau de la Sarthe, *L'art de connaître les hommes*, 6:229–30.

61. Ibid.

62. Ibid., 231–32.

63. Moreau de la Sarthe was one of the first French authors to explore a broader definition of *anthropology* as an empirical science. See his review of Pinel's *Traité medico-philosophique* in *La Décade philosophique* 29 (1801), 458–59; quoted in Sergio Moravia, *Beobachtende Vernunft: Philosophie und Anthropologie in der Aufklärung*. Frankfurt: Fischer, 1989 [1970], 64.

64. François Magendie, "Quelques idées générales sur les phénomènes particuliers aux corps vivans," *Bulletin des sciences médicales* 4, no. 24 (1809): 147.

65. See John Lesch, *Science and Medicine in France: The Emergence of Experimental Physiology 1790–1855* (Cambridge, MA: Harvard University Press, 1984); and Stephen L. Jacyna, "*Medical* Science and *Moral* Science: The Cultural Relations of Physiology in Restoration France," *History of Science* 25, no. 2 (1987): 111–46.

66. François Magendie, *Précis élémentaire de physiologie* (Paris: Méquignon-Marvis, 1816–1817); 2nd ed. (rev. and enl.), 1825; 3rd ed., 1833; 4th ed., 1836; 5th ed., 1838. An American edition appeared in 1822 as *A Summary of Physiology*, trans. John Revere (Baltimore: Edward J. Coale, 1822). Unless otherwise indicated, passages quoted here are newly translated from the French.

67. On the heterogenous nature of Magendies's *Précis élémentaire de physiologie* and its (partly speculative) refutations of Bichat, see Michael Gross, "The Lessened Locus of Feeling: A Transformation in French Physiology in the Early Nineteenth Century," *Journal of the History of Biology* 12, no. 2 (1979): 231–71.

68. See Magendie, *Précis élémentaire de physiologie* (1st ed.), 292. Significantly, the passage praising Barthez's theory of jumping is missing from the second edition onward (2nd ed., 1825, 331).

69. His thesis was presented to the Faculté de Médecine de Paris on January 29, 1820; François-Désiré Roulin, *Propositions sur les mouvements et les attitudes de l'homme* (Paris: Didot jeune, 1820).

70. François-Désiré Roulin, "Recherches théoriques et expérimentales sur le mécanisme des attitudes et des mouvemens de l'homme," *Journal de physiologie expérimentale* 1, no. 3 (1821): 235.

71. Ibid., 235–36.

72. Ibid., 228.

73. For some time, Roulin (who was also a protégé of Georges Cuvier) remained a faithful contributor to Magendie's journal. He eventually turned to natural history; he also wrote feuilleton articles for the *Revue des deux mondes*. For his biography, see Marguerite Combes, *Pauvre et aventureuse bourgeoisie: Roulin et ses amis* (Paris: J. Peyronnet, 1928).

74. Magendie, "Influence du cerveau sur les mouvements," in *Précis élémentaire de physiologie*, 2nd ed. (1825), 334–47.

75. Ibid., 337. See also Magendie, "Note sur les fonctions des corps striés et des

tubercules quadrijumeaux," *Journal de physiologie expérimentale et pathologique* 3, no. 4 (1823): 376–81.

76. Magendie, *Précis élémentaire de physiologie*, 2nd ed. (1825), 345.

77. See John Lesch, *Science and Medicine*; and Lesch, "The Paris Academy of Medicine and Experimental Science, 1820–1848," in *The Investigative Enterprise: Experimental Physiology in Nineteenth-Century Medicine*, ed. William Coleman and Frederic L. Holmes (Berkeley: University of California, 1988), 100–137.

78. For good reasons, Magendie's periodical was eventually renamed the *Journal de Physiologie expérimentale et pathologique*. On Magendie's "pathological physiology," see Lesch, *Science and Medicine*, 166–96.

79. Magendie, *Précis élémentaire de physiologie*, 2nd ed. (1825), 360.

80. Ibid., 361.

81. Ibid., 338–39.

82. Ibid., 339.

83. See Antoine de Montègre's extensive discussion of Saint-Médard's famous "convulsionnaires": "Convulsionnaires," in *Dictionnaire des sciences médicales* (Paris: Panckoucke, 1813), 6:210–38. Here Montègre also mentions the demonic possessions at Loudun: "it was here, to accomplish a work of iniquity and terrible vengeance, that a combination of the most disgusting ignorance, the most relentless fanaticism, and the most atrocious cruelty could be witnessed" (227). Esquirol devoted a long article to *démonomanie* (a term he coined), which was a variant of "religious melancholy": "Démonomanie," in *Dictionnaire des sciences médicales* (Paris: Panckoucke, 1814), 294–318.

84. On the "traitement moral," see Marcel Gauchet and Gladys Swain, *La pratique de l'esprit humain: L'institution asilaire et la révolution démocratique* (Paris: Gallimard, 1980); Jan Goldstein, *Console and Classify: The French Psychiatric Profession in the Nineteenth Century* (Cambridge: Cambridge University Press, 1987). On the many treatments for melancholy in this period, see Jean Starobinski, *Histoire du traitement de la mélancolie des origines à 1900* (Basel: Geigy, 1960).

85. Pierre-Nicolas Gerdy, *Physiologie médicale, didactique et critique*, vol. 1, pt. 1 (Paris: Crochard, 1832 [1830]), viii–ix.

86. Gerdy, *Essai de classification naturelle et d'analyse des phénomènes de la vie* (Paris: J.-B. Baillière, 1823 [1821]), vi–vii.

87. Gerdy, "Mémoire sur le mécanisme de la marche de l'homme," *Journal de Physiologie* 9 (1829), 1. We also know from a cover letter that Gerdy had lectured on physiology as an *aide d'anatomie* at the Faculté as early as 1817; see Gerdy, "[Lettre] à MM. les juges du concours de physiologie," May 30, 1831, 2, Bibliothèque interuniversitaire de santé, Paris. Gerdy's life and professional career are retraced in Paul Broca, *Eloge historique de P. N. Gerdy* (Paris: Bureau du Moniteur des Hôpitaux, 1856).

88. Gerdy, *Anatomie des formes extérieures du corps humain, appliquée à la peinture, à la scultpure et à la chirurgie* (Paris: Béchet jeune, 1829), 6; see also Gerdy, *Physiologie médicale*, vol. 1, pt. 1, 156–57.

89. Gerdy, *Anatomie*, ix.

90. The vivisection controversy was played out on the stage of the Paris Académie de médecine, which Gerdy joined in 1837. He viewed the many contradictions in the

experimental results reported by Magendie's students as contributions not to the history of nature but rather, as he mockingly remarked, to a "poetry of science." Gerdy, "Discussion sur les fonctions du système nerveux," *Bulletin de l'Académie Royale de Médecine* 3 (1838/39): 415. See the fairly one-sided account in Lesch, "Paris Academy of Medicine," 128–33; and (as a corrective) Paul Elliott, "Vivisection and the Emergence of Experimental Physiology in Nineteenth-Century France," in *Vivisection in Historical Perspective*, ed. Nicolaas Ruppke (London: Routledge, 1987), 48–77.

91. Gerdy, *Physiologie médicale*, vol. 1, pt. I, ciii.

92. Nicolas Philibert Adelon, *Physiologie de l'homme* (Paris: Compère jeune, 1823), 2:186.

93. Gerdy, *Physiologie médicale*, vol. 1, pt. 2, 495.

94. Adelon, *Physiologie*, vol. 2, 197.

95. Gerdy, *Physiologie médicale*, vol. 1, pt. 2, 572.

96. Ibid., 578.

97. Honoré de Balzac, *Théorie de la démarche*, in *La comédie humaine*, vol. 12 (Paris: Gallimard, 1981), 259–302.

98. See Moïse Le Yaouanc, *Nosographie de l'humanité balzacienne* (Paris: Maloine, 1959).

99. In this regard, many studies by Balzac scholars, as well as the commentary on the Pléiade edition of the *Comédie humaine*, led by Castex, have made peace with the contradictions in Balzac's oeuvre and sometimes even pushed them further. For a few starting points, see Madeleine Ambrière, "Balzac le chercheur d'absolu: Du coté de la science," in *Au soleil du romantisme*, 213–306, and *Balzac et la recherche de l'absolu* (Paris: Presses universitaires de France 1999).

100. For a nuanced analysis of the many contradictory connections between psychiatry and literature in Balzac's text, see Juan Rigoli, *Lire le délire*, 477–517.

101. Félix Davin, "Introduction par Félix Davin aux *Études philosophiques*," in Honoré de Balzac, *Œuvres completes* (Paris: Gallimard, 1979), 10:1211, 1213. "Social individualities" (*individualités sociales*) is a Balzac neologism meant to sound scientific.

102. Not until 1846 did Gerdy finally publish a *Physiologie philosophique des sensations et de l'intelligence fondée sur des recherches et des observations nouvelles et applications à la morale, à l'éducation, à la politique* (Paris: Labé).

103. Frédéric Dubois d'Amiens, "Physiologie médicale didactique et critique par M.P.N. Gerdy," *Revue médicale française et étrangère* 4 (1831): 445–67.

104. Walter Benjamin, *The Arcades Project*, trans. Howard Eiland (Cambridge, MA: Harvard University Press, 2002). Some recent studies of the *Physiologies* by literary historians have distanced themselves from Benjamin. See Martina Lauster, "Walter Benjamin's Myth of the Flâneur," in *Modern Language Review* 102 (2007), 139–56; *Sketches of the Nineteenth Century: European Journalism and Its Physiologies, 1830–50* (Palgrave: Macmillan, 2007); and Nathalie Preiss, *Les physiologies en France au XIXe siècle: Étude historique, littéraire et stylistique* (Mont-de-Marsan: Éditions InterUniversitaires, 1999).

105. See especially Philippe Buchez's theories in "De la physiologie," in *Le producteur* 3 (1826), 122–23, 264–80, 459–78; on Balzac and the Saint-Simonists, see Bruce

Tolley, "Balzac et les saint-simoniens," *L'année balzacienne* (1966), 49–66, and "Balzac et la doctrine saint-simonienne," in *L'année Balzacienne* (1973), 159–67.

106. Jean Anthèlme Brillat-Savarin: *Physiologie du goût, ou Méditations de gastronomie transcendante: Ouvrage théorique, historique et à l'ordre du jour, dédié aux gastronomes parisiens par un professeur, membre de plusieurs sociétés savants,* 3rd. ed. (Paris: A. Sautelet, 1829), 1:47.

107. Ibid., 101. For Brillat-Savarin and the gastronomic discourse in France, see Pascal Ory, *Le discours gastronomique français* (Paris: Gallimard, 1998), esp. 71–86.

108. This circumstance has led some literary scholars to view the *Théorie de la démarche* as a poststructuralist text avant la lettre. This reading seems questionable, however, at least when the text is understood within the historical context of scientific theories of walking, a context that is absent from the otherwise often excellent commentary to the Pléiade edition.

109. Balzac, *Théorie de la démarche,* 278.

110. Ibid., 280.

111. Ibid., 289–90.

112. Ibid., 291.

113. See Balzac's letter of October 20, 1822 to his sister Laure Surville, in Honoré de Balzac, *Correspondance I (1809–1835)* (Paris: Gallimard, 2006), 141.

114. Similarly, Balzac's *Traité de la vie élégante* (1830), written for the periodical *La Mode*, reads like a Romantic variation on the social physiology of the working classes that Moreau had presented: Balzac, *La comédie humaine* (Paris: Gallimard, 1981), 12:211–57. Considering how closely Balzac follows Moreau, the familiar interpretation of Balzac's work as a larger, literary reworking of Lavater's physiognomy is not very convincing: for this kind of approach, see, for example, Hans Ludwig Scheel, "Balzac als Physiognomiker," *Archiv für das Studium der neueren Sprachen und Literaturen* 198 (1962): 227–44, and Christopher Rivers, "'L'homme hieroglyphié': Balzac, Physiognomy, and the Legible Body," in *Faces of Physiognomy,* ed. Ellis Shookman (Columbia, SC: Camden House, 1993), 144–60.

115. Balzac, *Théorie de la démarche,* 272.

116. Besides the discussion above, see especially the Baron Joseph-Marie Degérando's *Considération sur les diverses méthodes à suivre dans l'observation des peuples sauvages* (n.p., 1800); in English as *The Observation of Savage Peoples,* trans. F. C. T. Moore (London: Routledge & Kegan Paul, 1969). As the first and most important of a number of necessary observations, Degérando instructed traveling ethnographers to study empirically the signs, and especially the gestural language, used by the "Savages." See also Degérando, *Des signes, et de l'art de penser considérés dans leurs rapports mutuels,* 4 vols. (Paris, 1800), and chapter 5 in Sophia Rosenfeld, *A Revolution in Language: The Problem of Signs in Late Eighteenth-Century France* (Stanford, CA: Stanford University Press, 2001).

117. If we consider Balzac solely within the literary tradition of Louis Sebastian Mercier's *Tableaux de Paris*—which have been read as an archetypal mythology of everyday life for contemporary readers—then his references to natural history and physiology are reduced to the status of prestige-granting "metaphorical resources"; Karl-Heinz Stierle, "Epische Naivität und bürgerliche Welt: Zur narrativen Struktur

im Erzählwerk Balzacs," in *Honoré de Balzac*, ed. Hans-Ulrich Gumbrecht, Karl-Heinz Stierle, and Rainer Warning (Munich: Fink, 1980), 182. Then the *Théorie de la démarche* appears as an ironic "carnivalization" of scholarly discourse in which a "semiotic sensorium for a phenomenology of the metropolis" manifests itself. Stierle, *Der Mythos von Paris: Zeichen und Bewußtsein der Stadt* (Munich: Hanser, 1993), 359.

118. Critical studies of the later genres of literary physiologies—most of which refer to Walter Benjamin's Paris studies and his theory of the *flâneur*—have tended to interpret Balzac's "physiological" essays in this particular sense. See especially Richard Sieburth, "Une idéologie du lisible: Le phénomène des physiologies," in *Romantisme* 47 (1985), 39–60, esp. 45. Attempts to correct this tendency can be found in Preiss, *Les physiologies*, and Lauster, *Sketches*.

119. Balzac, *Théorie de la démarche*, 277.

120. See Daston, "Empire of Observation," and *Eine kurze Geschichte der wissenschaftlichen Aufmerksamkeit* (Munich: Siemens Stiftung, 2001).

121. Elsewhere, Balzac depicts the splitting of the Institut de France into separate academies as a collapse of the unity of the sciences (in *Louis Lambert* and *Séraphita*, especially). The critique is reiterated verbatim by Geoffroy Saint-Hilaire in his *Notions synthetiques, historiques et physiologiques de philosophie naturelle*. See Madeleine Ambrière, "Les enchanteurs de la science: Balzac et les Messieurs du Muséum (1833)," in *Au soleil du romantisme*, 230–48.

122. Balzac, *La peau de chagrin*, in *La comédie humaine* (Paris: Gallimard, 1979), 10:243–44.

123. Balzac, *Théorie*, 226.

124. The figurative meaning became established in the late seventeenth century. See *Dictionnaire historique de la langue française*, ed. Alain Rey et al. (Paris: Le Robert, 1994), s.v. "marche."

125. Balzac, *Théorie de la démarche*, 265.

126. As in the *Physiologie du mariage* ("Physiologie, que me veux-tu?"), *Œuvres completes*, vol. 11, 865-1205 (Paris: Gallimard, 1977), theory here is personified as a capricious "courtesan" (266, 274).

127. Balzac, *Théorie de la démarche*, 270. This argument has been repeatedly linked to a general energetics of human life that permeates the *Human Comedy* and is thought to have its origins in Balzac's failed project of a "Essai sur les forces humaines." A very clear treatment is in Ernst Robert Curtius, *Balzac* (Frankfurt: Fischer, 1985 [1923]), who views the *Human Comedy* as the "démonstration de tout un système" and the energetics of life as a "fully coherent organic whole" (63). Balzac editor Madeleine Ambrière offers a similar view in her *Balzac et la recherche de l'absolu*. Here it seems more important to point out the disjointedness of these ideas on energetics, which Balzac puts to different tasks in different texts. In the *Théorie de la démarche*, for instance, Balzac's theory of movement builds directly on theories of willpower developed by his character Louis Lambert within his system of occult science (Balzac invented Lambert around this time). For a more detailed discussion, see chapter 5 in Jean Starobinski, *Action and Reaction. The Life and Adventures of a Couple* (New York: Zone, 2003).

128. Balzac, *Théorie de la démarche*, 270.

129. Ibid., 263.

130. Ibid., 273. Although Balzac says that Borell's text was first brought to his attention by the astronomer Félix Savary, and although he describes how he procured his copy, the fact that he consistently gets the title wrong (*De actu animalium*) makes it more likely that he was relying on other sources. Aside from the articles already mentioned, these might have included, for example, the extensive article "Marche" by Rullier from the *Dictionnaire des sciences médicales*, as it, too, reiterates Borelli's most important theories.

131. Balzac, *Théorie de la démarche*, 273.

132. Ibid., 274.

133. Ibid., 294.

134. Ibid., 285.

135. Ibid., 302.

CHAPTER 3

1. Balzac, *Théorie de la démarche*, in *La comédie humaine*, vol. 12 (Paris: Gallimard, 1981), 261.

2. Wilhelm Weber and Eduard Weber, *Mechanik der menschlichen Gehwerkzeuge: Eine anatomisch-physiologische Untersuchung*, in *Wilhelm Weber's Werke*, ed. Friedrich Merkel and Otto Fischer (Berlin: Julius Springer, 1894 [1836]), 6:27. An English translation was published as *Mechanics of the Human Walking Apparatus*, trans. P. Maquet and R. Furlong (Berlin: Springer, 1992). The excerpts presented here have been newly translated from the German.

3. The term is clearly modeled on *Sehwerkzeuge*, a word for the eyes. It does not appear in dictionaries or scientific literature of the time. It was employed in a casual way in 1831 by Prinz Maximilian zu Wied in his descriptions of bird legs in *Beiträge zur Naturgeschichte von Brasilien* (Weimar: Landes-Industrie-Comptoirs, 1831), vol. 3. pt. 2, 1051.

4. *Conversations-Lexikon der Gegenwart*, s.v. "Weber (Wilhelm Eduard)" (Leipzig: Brockhaus, 1841), vol. 4, pt. 2, 351.

5. This status can be traced from such linear histories of progress as Friedrich Wilhelm Bessel, *Populäre Vorlesungen über wissenschaftliche Gegenstände* (Hamburg: Perthes-Besser & Mauke, 1848), 326–86, via the heroic depiction of Gauß's and Weber's experiments in Heinrich von Treitschke, *Deutsche Geschichte im Neunzehnten Jahrhundert* (Leipzig: Hendel, 1927 [1879]), 4:575, until Daniel Kehlmann's popular-scientific novel *Die Vermessung der Welt* (Reinbek: Rowohlt, 2005).

6. Although its theory of walking has long been refuted, the Weber study is ritually, almost hagiographically cited as one of the founding texts of modern biomechanics. See, for example, J. P. Paul: "History and Fundamentals of Gait Analysis," in *Bio-Medical Materials and Engineering* 8, no. 3/4 (1998): 123–35. Historians of science, in contrast, have not addressed the work in any detail. In one paper (which, however, addresses scientific practice only briefly) Mary Mosher Flesher reads the Webers' research on human locomotion as an implicit critique of the Prussian military drill; see "Repetitive Order and the Human Walking Apparatus: Prussian Military Science

versus the Webers' Locomotion Research," *Annals of Science* 54, no. 5 (1997): 463–87. In view of the late eighteenth-century attempts by the military to measure precisely and regulate infantry steps as discussed in chapter 1, however, this interpretation appears questionable.

7. Weber and Weber, *Mechanik der menschlichen Gehwerkzeuge,* 269.

8. Ibid., 9.

9. Ibid.

10. Ibid., 178n1.

11. For information on the Weber brothers (general treatments, as little is known about their early lives), see Robert Knott, "Weber, Wilhelm," *Allgemeine Deutsche Biographie* 41 (1896): 358–61; Julius Leopold Pagel, "Weber, Eduard Friedrich," *Allgemeine Deutsche Biographie* 41 (1896): 287; and Heinrich Weber, *Wilhelm Weber: Eine Lebensskizze* (Breslau: E. Trewendt, 1893).

12. See Christiane Eisenberg, *"English Sports" und Deutsche Bürger: Eine Gesellschaftsgeschichte 1800–1939,* (Paderborn: Ferdinand Schöningh, 1999), chap. 2. Eisenberg details the differences between the gymnastics of late-Enlightenment philanthropinist educators and the gymnastics movement (*Turnbewegung*), including the paramilitary character of the latter.

13. Friedrich Ludwig Jahn and Ernst Wilhelm Eiselen, *Die Deutsche Turnkunst zur Einrichtung der Turnplätze* (Berlin: printed by the author, 1816), 3.

14. Ibid., 4. In England, a culture of competitive sports was established earlier. Here, "celebrated pedestrians" competed with each other, and in contrast to Germany, their achievements were measured more precisely. See Walter Thom, *Pedestrianism; or, An Account of the Performances of Celebrated Pedestrians during the Last and Present Century: With a Full Narrative of Captain Barclay's Public and Private Matches; and an Essay on Training* (Aberdeen: A. Brown and F. Frost, 1813).

15. Friedrich Ludwig Jahn, *Deutsche Turnkunst: Zum zweiten Male und sehr vermehrt herausgegeben* (Berlin: Reimer, 1847), 96. Authors (including Jahn's assistant Eiselen) writing essays on a theory of fencing during the same period adopted Weber's *Mechanik* for that purpose as well. Thus, Johann Wilhelm Roux, for instance, thought "that a comprehensive mechanics of thrust- and cut-fencing ought to be developed by analogy": Johann Wilhelm Roux, *Anweisung zum Hiebfechten mit geraden und krummen Klingen* (Jena: Friedrich Mauke, 1840), 29.

16. Franz Joseph Ritter von Gerstner, *Handbuch der Mechanik, aufgesetzt, mit Beiträgen von neuern englischen Konstruktionen vermehrt und herausgegeben von Franz Anton Ritter von Gerstner,* 2nd ed., vol. 1 (Prague: Johann Spurny, 1833).

17. Widely disseminated and popular texts included the lectures by Charles Dupin, a professor at the Conservatoire des arts et métiers. The lectures emphasized the importance of using correctly calculated centers of gravity in depictions of walking in the visual and performing arts, in determining load and alignment for marching infantry, and in machine construction. Carl Dupin, *Geometrie und Mechanik der Künste und Handwerke und der schönen Künste* (Paris, Straßburg: Levrault, 1826), vol. 2, esp. 53–100; published in French as *Géométrie et mécanique des arts et métiers et des Beaux-Arts: Cours normal à l'usage des artistes et des ouvriers, des sous-chefs et des chefs d'ateliers et de manufactures,* 3 vols. (Paris: Bachelier, 1826–1828). Weber and Weber,

Mechanik der menschlichen Gehwerkzeuge, 3–5. See also Wilhelm Weber's letter of April 16, 1834, to the Göttingen astronomer Johann Franz Encke, in which he emphasized the interest of the text for "educated military men," Nachlaß Gauß, Göttingen University Library.

18. Jean Théophile Desaguliers, *Cours de physique expérimentale* (Paris: Rollin, 1751); Charles de Coulomb, "Résultat de plusieurs expériences destinées à déterminer la quantité d'action que les hommes peuvent fournir par leur travail journalier, suivant les différentes manières dont ils emploient leurs forces," in *Théorie des machines simples, en ayant égard au frottement de leurs parties et à la roideur des cordages* (Paris: Bachelier, 1821), 255–97. On the origins of the science of work (*Arbeitswissenschaft*), see especially François Vatin, *Le travail et ses valeurs* (Paris: Albin Michel, 2004), which also offers a critical revision of Rabinbach's arguments in *The Human Motor*.

19. We can assume that the work of Gerstner, a mathematician who taught in Prague, was well known to the young Wilhelm Weber, who had been appointed professor of the theory of machines in Halle. The latter's *Wellenlehre* (1826), written with his brother Ernst Heinrich, discusses Gerstner's wave theory. In their *Mechanik der menschlichen Gehwerkzeuge*, however, the only recent literature that the Webers cite is the second volume of Siméon Denis Poisson's *Traité de mécanique* (Paris 1833).

20. Coulomb's *Théorie des machines simples* was first published in 1781 (see also Coulomb, *Théorie des machines simples* (1821), 1–186).

21. Wolfgang Schivelbusch, *The Railway Journey: The Industrialization of Time and Space in the Nineteenth Century* (Berkeley: University of California Press, 1986 [German 1st ed. 1977]), 16–33. In his vivid depiction, Schivelbusch draws attention to the definition of the railway line as an ideal, frictionless street.

22. James Adamson, *Sketches of Our Information as to Railroads* (Newcastle: Edward Walker, 1826), 51–52; also quoted in Schivelbusch, *Railway Journey*, 8–9.

23. Thomas Gray, for instance, one of the most active railway advocates, presented the railway as the perfect solution for overcoming the dangers of traveling by coach (on which, see chap. 1, above): "The numerous dangers and inconveniencies [*sic*] to which the present coach system is obnoxious, (such as the intractableness of horses, the imprudence of drivers, cruelty to animals, the dust and ruggedness of roads, &c.), would not be encountered on the rail-way, whose solid basis and peculiar construction render it impossible for any vehicle to be upset, or driven out of its course, the rail being convex and the rim of the carriage wheel concave; and as the rail-way must also be perfectly level and smooth, no danger could be apprehended from the increased speed, for mechanic power is uniform and regular, whilst horse-power, as we all very well know, is quite the reverse." Thomas Gray, *Observations on a General Iron Railway, or Land Steam-Conveyance: To Supersede the Necessity of Horses in All Public Vehicles; Showing Its Vast Superiority In Every Respect, over All the Present Pitiful Methods of Conveyance by Turnpike Roads, Canals, and Coasting-Traders*, 5th ed. (London: Baldwin, Cradock, and Joy, 1825), 71.

24. Weber and Weber, *Mechanik der menschlichen Gehwerkzeuge*, 4.

25. See Alexander Gordon, *Historische und practische Abhandlung über Fortbewegung ohne Thierkraft mittelst Dampfwagen auf gewöhnlichen Landstraßen*

(Weimar: Landes-Industrie-Comptoirs, 1833 [1832]); Ludwig Newhouse, *Über Chaussee-Dampfwagen, statt Eisenbahnen mit Dampfwagen in Deutschland* (Mannheim: Hoff, 1834); review in the gratis supplement to the January 1835 issue of the *Vaterländischer Berichte für das Großherzogtum Hessen und die übrigen Staaten des deutschen Handelsvereins* 1 (1835): 12–14; Anon., "Über Chausseedampfwagen und Pferdeeisen-bahnen," *Polytechnisches Journal* 52 (1834): 401–8; and Anon., "Über Dampfwagen," *Bayerische National-Zeitung*, October 4, 1834, 1092–94; October 5, 1834, 1096–97.

26. Anon., "Ueber Chausseedampfwagen," 401.

27. J. W. Schmitz, "Wahrscheinliche Resultate der Dampfwagen auf gewöhnlichen Wegen," *Der Sammler: Ein Unterhaltungsblatt* 85 (July 17, 1834): 340–41.

28. Joseph Ritter von Baader, *Die Unmöglichkeit, Dampfwagen auf gewöhnlichen Straßen mit Vortheil als allgemeines Transportmittel einzuführen, und die Ungereimt-heit aller Projekte, die Eisenbahnen dadurch entbehrlich zu machen* (Nürnberg: Riegel und Wießner, 1835), 23.

29. Anon., "Ueber selbstfahrende Fuhrwerke," *Polytechnisches Journal* 55 (1835): 12. See also Victor Mekarsky Edler von Menk, *Das Eisenbahnwesen nach allen Bezie-hungen kritisch beleuchtet: Für den Gebildeten jeden Standes und ein vollständiges Handbuch für Eisenbahn-Comittéen, Privat-Unternehmer, Mit-Interessenten, Archi-tekten, Ingenieurs und Mechaniker* (Vienna: Franz Tendler, 1837), 31–37; and Nicolaus N. W. Meissner, *Geschichte und erklärende Beschreibung der Dampfmaschinen, Dampfschiffe und Eisenbahnen nebst einer Erläuterung der Natur der Wasserdämpfe und der dabei vorkommenden Kunstausdrücke für diejenigen, denen Kenntnisse in Mechanik, Mathematik und Physik fehlen* (Leipzig: G. Fleischer, 1839), 145–51.

30. Weber and Weber, *Mechanik der menschlichen Gehwerkzeuge*, 5.

31. Moving powers included, according to the theory of machines from around 1835, not only the "powers of humans and some animals that have been made subservient to us" but also weights, springs, water, wind, air and steam. See, for example, Chris-toph Bernoulli, *Elementarisches Handbuch der industriellen Physik, Mechanik und Hydraulik* (Stuttgart: Cottasche Buchhandlung, 1835), 2:9.

32. Weber and Weber, *Mechanik der menschlichen Gehwerkzeuge*, 5.

33. Ibid., 13.

34. From their correspondence, we learn that the Webers repeated some of their ex-periments while traveling. See, for example, Wilhelm Weber's letter to C. A. Steintheil, October 24, 1836, Archive Deutsches Museum Munich.

35. Weber and Weber, *Mechanik der menschlichen Gehwerkzeuge*, 156.

36. Ibid., 299.

37. For the earlier military research on gait, see chapter 1. For the more general con-text in the natural sciences, see Christoph Hoffmann, *Unter Beobachtung: Naturfor-schung in der Zeit der Sinnesapparate* (Göttingen: Wallstein, 2006), 221–31.

38. See Wilhelm Weber's letter of April 16, 1834, to Encke in Göttingen: "In Berlin, you ever so kindly remembered my and my brother's intense occupation with walking and running and are sending me the excellent observations by Mr. Gersdorf. It is very desirable to get reliable information from that side; because we believe that, even if no direct practical benefit might be gained [from our work], an interest for the subject of our work nevertheless could be instilled, especially and most of all among educated

military men, but only if the best observations actually made in the army can be used in this work. However, we are having difficulty in finding a good source to fully share these experiences with us." (Nachlaß Gauß, Göttingen University Library). See also Weber, *Wilhelm Weber*, 42.

39. Carl von Clausewitz, *Vom Kriege: Hinterlassenes Werk*, ed. Marie von Clausewitz (Berlin: Ferdinand Dümmler, 1832), 1:93; English translation from Clausewitz, *On War*, trans. and ed. Michael Eliot Howard and Peter Paret (Princeton, NJ: Princeton University Press, 1989), 120.

40. Such attempts were made, for example, during the German campaign by Hans von Bülow. Bülow tried to explain tactics and strategy using a mathematical theory of operational lines in *Lehrsätze des neuern Krieges oder reine und angewandte Strategie aus dem Geist des neuern Kriegssystems hergeleitet* (Berlin: Heinrich Frölich, 1805). In contrast to enterprises that understood military operations exclusively as a play of physical-mechanical masses stood the idea of a "national army" supported by "enthusiasm" and "military spirit," which Scharnhorst successfully propagated in Prussia. See Heinz Stübig, "Berenhorst, Bülow und Scharnhorst als Kritiker des preußischen Heeres der nachfriderizianischen Epoche," in *Die preußische Armee: Zwischen Ancien Régime und Reichsgründung*, ed. Peter Baumgart, Bernhard R. Kroener, and Heinz Stübig (Paderborn: Ferdinand Schönigh, 2008), 107–20.

41. Clausewitz, *Vom Kriege*, 93; English translation from Clausewitz, *On War*, 119.

42. Clausewitz, *Vom Kriege*, 92; English translation from Clausewitz, *On War*, 119.

43. Clausewitz, *Vom Kriege*, 94–95; English translation from Clausewitz, *On War*, 120.

44. See above, pp. 30–31.

45. The pendulum was regarded as a model for precise mathematical measurements not least thanks to refinements made by Bessel, who used the instrument when tasked by the government in 1833–1839 with defining a primary standard of measurement for the Prussian foot. See Kathryn Olesko, "The Meaning of Precision: The Exact Sensibility in Early Nineteenth-Century Germany," in *The Values of Precision*, ed. M. Norton Wise (Princeton, NJ: Princeton University Press, 1995), 103–34.

46. Eduard Weber, "Einige Bemerkungen über die Mechanik der Gelenke, insbesondere über die Kraft, durch welche der Schenkelkopf in der Pfanne erhalten wird; ein Vortrag gehalten vor der Versammlung der deutschen Naturforscher zu Bonn am 23. September," *Archiv für Anatomie, Physiologie und wissenschaftliche Medicin* 1836:54–59.

47. Weber and Weber, *Mechanik der menschlichen Gehwerkzeuge*, 108.

48. Letter by Alexander von Humboldt to Altenstein, October 30, 1837, in Alexander von Humboldt, *Alexander von Humboldt: Vier Jahrzehnte Wissenschaftsförderung; Briefe an das preußische Kultusministerium 1818–1859* ed. Kurt Biermann (Berlin: Akademie, 1985), 81.

49. Wilhelm and Eduard Weber, "Ueber die Mechanik der menschlichen Gehwerkzeuge, nebst der Beschreibung eines Versuchs über das Herausfallen des Schenkelkopfs aus der Pfanne im luftverdünnten Raum," *Annalen der Physik und Chemie* 40, no. 1 (1837): 8. Alexander von Humboldt, *Über einen Versuch den Gipfel des Chimborazo zu ersteigen: Mit dem vollständigen Text des Tagebuches "Reise zum Chimborazo,"*

ed. Oliver Lubrich and Ottmar Ette (Frankfurt: Eichborn, 2006), 141. First published as "Ueber zwei Versuche den Chimborazo zu besteigen," in *Jahrbuch für 1837*, ed. H. C. Schumacher (Stuttgart: Cotta, 1837), 176–206; 193–94.

50. Weber and Weber, "Ueber die Mechanik der menschlichen Gehwerkzeuge," 10.

51. Ibid., 13.

52. See Rafael Mandressi, *Le regard de l'anatomiste: Dissections et invention du corps en Occident* (Paris: Seuil, 2003).

53. See Pierre-Nicolas Gerdy, *Anatomie des formes extérieures du corps humain, appliquée à la peinture, à la scultpure et à la chirurgie* (Paris: Béchet jeune, 1829), and above, pp. 59–62.

54. Weber and Weber, *Mechanik der menschlichen Gehwerkzeuge*, (1836), viii–ix.

55. Ibid., 17.

56. Ibid., 3. The original illustration is in Bernhard Siegfried Albinus, *Tabulae sceleti et musculorum corporis humani* (Lugduni Batavorum: Johannem & Hermannum Verbeek, 1747). The Webers' zeal for improvement was limited to representations of the male skeleton; women were excluded from their study.

57. See Lorraine Daston and Peter Galison, *Objectivity* (New York: Zone Books, 2007), 70–71; Sachiko Kusukawa, "The Uses of Pictures in the Formation of Learned Knowledge: The Cases of Leonhard Fuchs and Andreas Vesalius," in *Transmitting Knowledge: Words, Images, and Instruments in Early Modern Europe*, ed. Sachiko Kusukawa and Ian MacLean (Oxford: Oxford University Press, 2006) 73–96; and Irmgard Müller and Daniela Watzke, "Weil also die beste Abbildung [. . .] immer ein dürftiges Gleichnis bleibt: Zu den Visualisierungsverfahren im 18. Jahrhundert," in *Anatomie und anatomische Sammlungen im 18. Jahrhundert*, ed. Rüdiger Schultka and Josef N. Neumann (Berlin: Hopf, 2007) 223–50.

58. For a reconstruction of this method and the collaboration between Albinus and Wandelaar, see Hendrik Punt, *Bernard Siegfried Albinus (1697–1770): On "Human Nature"; Anatomical and Physiological Ideas in Eighteenth Century Leiden* (Amsterdam: B. M. Israël, 1983). For a more recent discussion, see Andrew Cunningham: *The Anatomist Anatomis'd: An Experimental Discipline in Enlightenment Europe*, (Farnham: Ashgate, 2010), 253–61.

59. Weber and Weber, *Mechanik der menschlichen Gehwerkzeuge*, 6.

60. See Dupin, *Geometrie und Mechanik der Künste*, chap. 3.

61. Weber and Weber, *Mechanik der menschlichen Gehwerkzeuge*, 299.

62. Ibid., 10–11.

63. Ibid., 62.

64. Ibid., 13.

65. Ibid., 32–33.

66. For the Webers' discussion of unconscious movements in walking and running and their rejection of self-observation as a method, see ibid., 10.

67. The theory appears in its most reduced form in the pictures on the final plate. Here, the position of the walking body is reduced to a set of points, and its motion is depicted as a succession of triangles. Variations in gait and speed are entered as graphs in a system of coordinates. In these figures, the real path traveled in the experiments is lifted into a geometrical space and translated into a graph. The figures are included

as a means of visualizing the measurements of stride duration and length recorded in a number of tables. Confined to the final table, however, they make much less of an impression than the bone prints and the sketches of the walking skeletons.

68. Weber and Weber, "Mechanik der menschlichen Gehwerkzeuge," 238.

69. See Jonathan Crary, *Techniques of the Observer* (Cambridge, MA: MIT Press, 1990), 105–12, and Maurice Dorikens, ed., *Joseph Plateau: 1801–1883: Leven tussen Kunst en Wetenschap* (Gent: Provincie Oost-Vlaanderen, 2001).

70. Simon Stampfer: *Die stroboskopischen Scheiben oder optischen Zauberscheiben* (Vienna: Trentsensky & Vieweg, 1833), 10.

71. Ibid., 11.

72. William George Horner, "On the Properties of the Daedaleum, a New Instrument of Optical Illusion," *Philosophical Magazine* 4 (1834), 36–41.

73. Weber and Weber, *Mechanik der menschlichen Gehwerkzeuge*, 304.

74. This advantage is mentioned by Horner: "no necessity exists in this case for bringing the eye near the apparatus, but rather the contrary, and the machine when revolving has all the effect of transparency, the phaenomenon may be displayed with full effect to a numerous audience." Horner, "On the properties of the Daedaleum," 37.

75. Weber and Weber, *Mechanik der menschlichen Gehwerkzeuge*, 295.

76. Johannes Müller, *Handbuch der Physiologie des Menschen für Vorlesungen*, 2nd ed. (Coblenz: Hölscher, 1840), 2:123.

77. "Meyer, Georg Hermann von," in Julius Leopold , ed., *Biographisches Lexikon hervorragender Ärzte des neunzehnten Jahrhunderts* (Berlin: Urban & Schwarzenberg, 1901), cols. 1126–27. Hermann Meyer, *Die richtige Gestalt der Schuhe: Eine Abhandlung aus der angewandten Anatomie für Aerzte und Laien geschrieben* (Zurich: Meyer & Zeller, 1858); *Die richtige Gestalt des menschlichen Körpers in ihrer Erhaltung und Ausbildung für das allgemeine Verständniß dargestellt* (Zurich: Meyer & Zeller, 1874).

78. Hermann Meyer, "Das aufrechte Stehen (Erster Beitrag zur Mechanik des menschlichen Knochengerüstes)," *Archiv für Anatomie, Physiologie und wissenschaftliche Medicin* 1853:11.

79. Ibid., 29.

80. Weber and Weber, *Mechanik der menschlichen Gehwerkzeuge*, 81.

81. Hermann Meyer, "Die Individualitäten des aufrechten Ganges (Vierter Beitrag zur Lehre von der Mechanik des menschlichen Knochengerüstes)," *Archiv für Anatomie, Physiologie und wissenschaftliche Medicin* 1853:548–49.

82. See also above, pp. 49–55.

83. Hermann Meyer, "Das aufrechte Gehen (Zweiter Beitrag zur Mechanik des menschlichen Knochengerüstes)," *Archiv für Anatomie, Physiologie und wissenschaftliche Medicin* 1853:365.

84. Ibid., 393.

85. Ibid.

86. Such a technique was, for instance, designed by Dr. Hugoulin of Toulon: M. Hugoulin,"Solidification des empreintes des pas sur les terrains les plus meubles en matière criminelle," *Annales d'hygiène publique et de médecine légale* 44 (1850): 429–32;"Reproduction des empreintes des pas, de coups de fusil, etc. sur la neige en matière criminelle," *Annales d'hygiène publique et de médecine légale*, n.s., 3 (1855):

207–12. In this case the forensic expert concerned himself not with the semiotics of different gaits but with the chemical procedure that guaranteed the best possible reproduction of a footprint. Usually, as also later, the preferred method for securing "foot imprints" was the plaster cast—as opposed to photography, for example—because this preserved their three-dimensionality. In a later, modified variant of this technique, the original imprint was even abandoned in favor of its reproduction; see Alphonse Jaumes, "Étude des procédés employés pour relever les empreintes sur le sol," *Annales d'hygiène publique et de médecine légale* 3, no. 3 (1880): 168–77.

87. Séverin Caussé, *Des empreintes sanglantes des pieds et de leur mode de mensuration* (Toulouse: Chauvin, 1853).

88. The two methods eventually proved compatible with one another, and both were extensively discussed in the literature. The British pathologist Charles Tidy, for instance, was dubious that Caussé's method of comparing bare footprints was suitable for the identification of criminals. Charles Tidy, *Legal Medicine*, vol. 1 (London: Smith, Elder, 1882). Tidy conducted a number of experiments to show that a footprint can be bigger or smaller than the corresponding foot or shoe, and he insisted that the structure of the soil in each case was the decisive factor.

89. See Régis Messac, *Le "detective novel" et l'influence de la pensée scientifique* (Paris: Encrage, 2011 [1929]),; and Carlo Ginzburg, "Clues: Roots of an Evidential Paradigm," in *Clues, Myths, and the Historical Method* (Baltimore: John Hopkins University Press, 1989), 96–125.

CHAPTER 4

1. [Guillaume-Benjamin] Duchenne de Boulogne, *Du second temps de la marche, suivie de quelques déuctions pratiques: Mémoire présenté à l'académie des sciences* (Paris: l'Union médicale, 1855), 6.

2. Ibid., 18.

3. [Wilhelm Weber and Eduard Weber], "Mécanique de la locomotion chez l'homme," in: *Encyclopédie anatomique II: Ostéologie, syndesmologie, et mécanique des organes locomoteurs*, trans. A.-J.-L. Jourdan (Paris: J. B. Ballière, 1843), 237–522. An early critique of the Webers' work can be found in the studies of Jacques Maissiat (*Études de physique animale* [Paris: Béthune et Plon, 1843]), who was an assistant to the zoologist Georges Duvernoy, Cuvier's nephew, at the Collège de France. While Maissiat's theory of locomotion bears a striking resemblance to the work of the Webers (which was known and discussed in Paris from at least 1837), he claimed rather implausibly to have arrived at the pendulum theory independently, through his own observations, and criticizes the many imprecisions in the Webers' measurements.

4. Felix Giraud-Teulon, *Principes de mécanique animale; ou, Étude de la locomotion chez l'homme et les animaux vertébrés* (Paris: J.-B. Ballière et fils, 1858), 221.

5. Felix Giraud-Teulon, "Locomotion," in *Dictionnaire encyclopédique des sciences médicales*, 2nd ser. (Paris: Asselin / Masson et fils, 1869), 2:786–87.

6. In contrast to most of the other observers of movement that we have encountered so far, over the last forty years Marey has become the subject of a wide-ranging body of specialized literature. This research has mostly focused on his biography and his

influence on the history of photography and film. Attempts at contextualization have remained the exception or have been selective and too schematic. See, for example, Rabinbach's attempt to portray Marey as an emblematic figure of "social modernism," serving the "politics of a state devoted to maximizing the economy of the body," in Anson Rabinbach, *The Human Motor: Energy, Fatigue, and the Origins of Modernity* (Berkeley: University of California Press, 1990), 119. Marey's own rhetoric has been uncritically adopted by many authors who rarely engage with the details of the practical functioning of his recording devices or the controversies they inspired. Despite their merits, this is true of François Dagognet's *Etienne-Jules Marey: A Passion for the Trace*, trans. Robert Galeta and Jeanine Herman (New York: Zone Books, 1992 [1987]) and Marta Braun's *Picturing Time: The Work of Etienne-Jules Marey (1830–1904)* (Chicago: University of Chicago Press, 1992). More recent studies by French film and media historians have added important source material but have not shifted away from the hagiographic topos of the Mareysian "revolution" in physiology. See Laurent Mannoni, *Etienne-Jules Marey: La mémoire de l'œil* (Milan: Mazzotta; Paris: Cinémathèque française, 1999) and Michel Frizot, *Etienne-Jules Marey: Chronophotographe* (Paris: Nathan, 2001).

7. Until the early twentieth century, urban transport technologies still ran largely on horsepower. Despite their great importance, horses have only recently emerged as a legitimate topic of study for historians. The most fervent plea for studying horses can be found in Daniel Roche, "Equestrian Culture in France from the Sixteenth to the Nineteenth Century," *Past and Present* 199 (May 2008): 113–45; see also his three-volume study *La culture équestre de l'Occident, XVIe–XIXe siècle: L'ombre du cheval*, 3 vols. (Paris: Fayard, 2008–2015). This recent turn toward the history of animals owes much to anthropological studies (see, e.g., Jean-Pierre Digard, *Une histoire du cheval*, 2nd ed. (Arles: Actes Sud, 2007). For the United States, see the recent study by Ann Norton Greene, *Horses at Work: Harnessing Power in Industrial America* (Cambridge, MA: Harvard University Press, 2008).

8. See, for example, Laurent Mannoni, *The Great Art of Light and Shadow: Archaeology of the Cinema*, trans. and ed. Richard Crangle (Exeter: University of Exeter Press, 2006), 299–363.

9. The far-reaching consequences of the controversy between d'Aure (1799–1863) and Baucher (1796–1873) have not yet been the subject of a detailed historical inquiry. For a social history of the circus, see Caroline Hodak, "Du théâtre équestre au cirque: 'une entreprise si éminemment nationale'; Commercialisation des loisirs, diffusion des savoirs et théâtralisation de l'histoire en France et en Angleterre (1760–1860)" (diss., École des hautes études en sciences sociales, 2004), 589–613. For the origins of the circus in England, see Marius Kwint, "The Legitimation of the Circus in Late Georgian England," *Past and Present* 174 (February 2002): 72–115, and "The Circus and Nature in Late Georgian England," in *Histories of Leisure: Leisure, Consumption, and Culture*, ed. Rudy Koshar (Oxford: Bloomsbury, 2002), 45–60.

10. Caroline Hodak, "Du spectacle militaire au théâtre équestre," in Roche, *Le cheval et la guerre: Du XVème au XXème siècle*, edited by Daniel Roche (Versailles: Association pour l'académie d'art équestre de Versailles, 2002), 367–77; "Créer du sensa-

tionnel: Spirale des effets et réalisme au sein du theatre équestre vers 1800," *Terrain* 46 (March 2006): 49–66.

11. François Baucher, *Dictionnaire raisonné de l'équitation,* 2nd ed. (Paris: F. Baucher, 1851 [1833]), 187.

12. After Baucher's methods were prohibited in the French army by an ordinance of the Ministry of War in 1853, Raabe (1810–1889) defended them in numerous writings. See, for example, Charles Raabe, *Examen du cours d'équitation de M. Aure, écuyer en chef de l'École de cavalerie, (Saumur 1852)* (Marseille: Marius Olive, 1854). Raabe's complete works were compiled by several of his students shortly before his death. See Étienne Barroil, *L'art équestre: Première partie; Iconographie des allures et des changements d'allures* (Paris: J. Rothschild, 1889); and Barroil, *L'art équestre: Deuxième partie; Dressage raisonné du cheval; avec une préface du Commandant Bonnal* (Paris: J. Rothschild, 1889).

13. Louis-Rupert Wachter, *Aperçus équestres au point de vue de la méthode Baucher* (Paris: Librairie militaire J. Dumaine, 1862), vi–vii. This also explains why many authors after Baucher chose the dictionary form. See Jules Pellier, *Le langage équestre,* 2nd ed. (Paris: Delagrave, 1900). The term *hippologie* to designate a special scientific branch appears first around 1855 in French. See Charles-Louis de Curnieu, *Leçons de science hippique générale; ou, Traité complet de l'art de connaître, de gouverner et d'élever le cheval,* 3 vols. (Paris: Librairie militaire J. Dumaine, 1855–1860).

14. The mechanistic view of horses as "living machines" gained widespread acceptance after Watt's invention of the steam engine and the ensuing attempts to measure "horsepower." The expression is taken up, rather uncritically, in a recent study focusing on the use of horses in the United States: Clay MacShane and Joel A. Tarr, eds., *The Horse in the City: Living Machines in the Nineteenth Century* (Baltimore: Johns Hopkins University Press, 2007).

15. Charles Raabe, *Examen du Bauchérisme réduit à sa plus simple expression, ou L'art de dresser les chevaux d'attelage, de dame, de promenade, de chasse, de course, d'escadron, de cirque, de tournoi, de carrousel* (Paris: Dumaine, 1857), 17.

16. Charles Raabe, "Lettre préface," in Barroil, *L'art équestre: Première partie.*

17. For the increasing physiological trend within pedagogy and gymnastics in the 1830s and 1840s, see the overview by Georges Vigarello, *Le corps redressé: Histoire d'un pouvoir pédagogique* (Paris: Armand Colin 2001 [1978]), 64–71.

18. "The play of forces of the animal machine is not easy to grasp: sometimes one needs the greatest sharpness of vision, the most rapid ability to comprehend, the most well-founded *raisonnement* in order to determine which body member took what role in an action that occurred with a speed and an instantaneity like that of thinking. Only by analyzing the phenomenon can one succeed, if not in explaining it, then at least in understanding it. Through the analysis one can finally comprehend it and achieve a formulation of its synthesis. Then the investigation has penetrated everything and solved all questions." Jacques Mignon, *Quelques réflexions sur la mécanique animale, appliquée au cheval* (Paris: Béchet jeune et Labé, 1841), 18.

19. Jules Lenoble Du Teil, *Étude sur la locomotion quadrupède* (Paris: J. Dumaine, 1873). The most well-known forerunner to this frequently used technique was the work

of Vincent and Goiffon, which aimed at facilitating the ideal depiction of the horse through the creation of an "artificial memory." Antoine-François Vincent and Georges-Claude Goiffon, *Mémoire artificielle des principes relatifs à la fidelle représentation des animaux, tant en peinture qu'en sculpture: Première partie concernant le cheval* (Alfort: Vincent and Goiffon, 1779).

20. Wachter, *Aperçus équestres*, 257.

21. See the discussion of the horse "types" in Louis Michel Morris, *Essai sur l'extérieur du cheval*, 2nd ed. (Paris: Bouchard-Huzard, 1857 [1835]), 1835, and also J.-P. Mégnin, *Essai sur les proportions du cheval et son anatomie extérieure comparée à celle de l'homme à l'usage des écuyers militaires ou des artistes* (Paris: Corréard, 1860).

22. Géricault and Delacroix, for instance, were among the many artists in regular attendance at the Cirque Olympique. On the role of the circus in equestrian painting, see Hodak, *Du théâtre équestre au cirque*, 601–3.

23. See Emile Duhousset, *Le cheval: Études sur les allures, l'extérieur et les proportions du cheval; Analyse de tableaux représentant des animaux; Dédié aux artistes* (Paris: Chasles, 1874), 62; also the significantly expanded version, *Le cheval: Allures, extérieur, proportions* (Paris: Morel, 1881), 112.

24. Théophile Gautier, "Artistes contemporains: Meissonier," *Gazette des Beaux-Arts*, 1st ser., May 12, 1862, 427–28.

25. Ibid., 422.

26. Octave Gréard, *Jean-Louis-Ernest Meissonier: Ses souvenirs—ses entretiens* (Paris: Hachette, 1897), 73. For a discussion of the painter's self-understanding as a scientist and the consequences thereof, see the excellent study by Marc J. Gotlieb, *The Plight of Emulation: Ernest Meissonier and French Salon Painting* (Princeton, NJ: Princeton University Press, 1996).

27. See the descriptions provided by Meissonier's son in François Thiébault-Sisson, "Meissonnier [sic]: Ses procédés de travail," *Le Temps*, November 6, 1895.

28. See Arthur Hartley Saxon, *Enter Foot and Horse: A History of Hippodrama in England and France* (New Haven: Yale University Press, 1968).

29. Thiébault-Sisson, "Meissonnier." Other accounts, for example that of Gérome, mention a "fauteuil roulant": see Jean-Louis Gérome, "Notes & Fragments des Souvenirs inédits du Maître," *Les arts* (1904): 30.

30. Pierre Véron, *Les coulisses artistiques* (Paris: Dentu, 1876), 235.

31. Étienne-Jules Marey, "Moteurs animés," *La Nature*, September 28, 1878, 273.

32. Étienne-Jules Marey, *La méthode graphique* (Paris: Masson, 1878), iii.

33. Ibid., iv. As this passage demonstrates, Marey made use of a very wide-ranging notion of graphics.

34. Ibid., xix.

35. See Soraya de Chadarevian, "Graphical Method and Discipline: Self-Recording Instruments in Nineteenth-Century Physiology," *Studies in History and Philosophy of Science* 24 (1993): 267–91. On Helmholtz and Ludwig, see Frederic L. Holmes and Kathryn M. Olesko, "The Images of Precision: Helmholtz and the Graphical Method in Physiology," in *The Values of Precision*, ed. M. Norton Wise (Princeton NJ: Princeton University Press, 1995), 198–221; Robert Brain, "The Graphic Method: Inscription,

NOTES TO PAGES 112–115

Visualization, and Measurement in Nineteenth-Century Science and Culture," (PhD diss., University of California, Los Angeles, 1996).

36. Marey, *La méthode graphique*, vi.

37. In many ways, Marey's strategies resemble the approach Bruno Latour identified in Louis Pasteur's bacteriological laboratory: see Latour, *The Pasteurization of France*, trans. Alan Sheridan and John Law (Cambridge, MA: Harvard University Press, 1988). What is more, Latour's own use of "inscription" can be traced back to Marey. In other respects, too, Marey's work can be read as a model for Latour's approach to modern science. See in particular Latour, "Drawing Things Together," in *Representation in Scientific Practice*, ed. Michael Lynch and Steve Woolgar (Cambridge, MA: MIT Press, 1990), 19–68.

38. Étienne-Jules Marey, *La machine animale: Locomotion aérienne et terrestre* (Paris: Librairie Germer Ballière, 1873). An English translation appeared the very next year in the International Scientific Series: *Animal Mechanism: a Treatise on Terrestrial and Aerial Locomotion* (London: King, Bradbury, Agnew, 1874 ; New York: Appleton, 1879). Since this translation is sometimes inaccurate, I have modified them or made new translations here.

39. Gaston Carlet, *Essai expérimental sur la locomotion humaine: Étude de la marche* (Paris: Masson, 1872).

40. Duchenne had included his earlier refutation of Weberian theory in his *Physiologie des mouvements* (Paris: J. B. Baillière et fils, 1867), 761–63. English translation taken from *Physiology of Motion*, trans. Emanuel B. Kaplan (Philadelphia: J. B. Lippincott, 1949), 550.

41. Carlet, *Essai*, 34n1.

42. This expression was applied not only to various animal species, but also to all sorts of vehicles and mobile machines. Even though it appears that Marey began using this concept relatively late, the systematic-comparative approach played a role in his work from the beginning. See Marey, "Moteurs animés."

43. Marey, *La machine animale*, 146. Marey does not give an exact citation, but the reference is clearly to Antoine Dugès' *Traité de physiologie comparée de l'homme et des animaux*, vol. 2 (Montpellier / Paris: Castel, 1838).

44. Both Carlet and Marey use the term *manège* to describe this circular path, which was also used in the study of bird flight. See Marey, La machine animale, 257–58; Animal Mechanism, 248–49.

45. Marey, *La machine animale*, 127.

46. "Compte-rendu de la séance du 27 juillet 1882," *Journal officiel, débats parlementaires*, pp. 1460–62.

47. Étienne-Jules Marey "La station physiologique," *La Nature*, September 8, 1883, 227.

48. Marey thus offers an interesting variation on the articulation of the lab-field border, a feature of the life sciences since the second half of the nineteenth century. See Robert Kohler, *Landscapes and Labscapes: Exploring the Lab-Field Border in Biology* (Chicago: University of Chicago Press, 2002), and for a more recent discussion, "Lab History," *Isis* 99 (2008): 761–68, and Graeme Gooday, "Placing or Replacing the Laboratory in the History of Science?," *Isis* 99 (2008): 783–95. A more detailed reconstruction

and analysis of the work done at the Station physiologique has not been undertaken so far. The major elements for such a study are preserved at the Musée Marey in Beaune, the Collège de France, and the Bibliothèque nationale in Paris. The most important set of published sources is in Thierry Lefebvre, Jacques Malthête, and Laurent Mannoni, eds., *Lettres d'Etienne-Jules Marey à Georges Demenÿ 1880–1894* (Paris: Association française de recherche sur l'histoire du cinéma, 1999).

49. See Françoise Forster-Hahn, "Marey, Muybridge and Meissonier: The Study of Movement in Science in Art," in *Eadweard Muybridge: The Stanford Years, 1872–1882* (Stanford: Stanford University Museum of Art, 1972), 85–109.

50. Marey, "La station physiologique," 230.

51. One of the most famous examples from the art world is the surrealist collages in *La femme 100 Têtes* (1929) by Max Ernst, in which the illustrations from *La Nature* are transformed using other elements. The Station physiologique can thus count as one of the most emblematic instances of "mechanical objectivity." For this notion, see Lorraine Daston and Peter Galison, "The Image of Objectivity," *Representations* 40 (1992): 81–128, which opens with Marey's case, and their more recent *Objectivity* (New York: Zone Books, 2007), from which the French physiologist is entirely absent.

52. See John Douard, who all too hastily assumes the persuasiveness of Marey's "visual rhetoric" in "E.-J. Marey's Visual Rhetoric and the Graphic Decomposition of the Body," *Studies in the History and Philosophy of Science* 26, no. 2 (1995): 175–204.

53. See the letter from Marey to his mother of February 3, 1882, printed in Marey, *Le mouvement* (Nimes: J. Chambon, 1994), 319.

54. Étienne-Jules Marey, "Études pratiques sur la marche de l'homme," *La Nature*, January 24, 1885, 119. That this ambition was thoroughly Marey's own is demonstrated by his use of chronophotography to analyze modern competitive athletics, which continued after his falling-out with Demenÿ and his withdrawal from the Station in 1894. See Marey's later study, "La chronophotographie et les sports athlétiques," *La Nature*, April 13, 1901, 310–15.

55. Gabriel Colin (1825–1896) had in 1871–1873 published his *Traité de physiologie comparée des animaux* (Paris: Baillière et fils, 1856) in a second edition (the first dated in fact from 1854–56). Thus, it is quite apparent that Marey's project was direct competition with his own.

56. "Even if everyone is in a position to record graphs, not everyone is in a position to interpret them. The method deceives even those who believe themselves in their scholarly fashion to be making use of it." Gabriel Colin, discussion of Marey's lecture in "Sur l'importance au point de vue médical des signes extérieurs des fonctions de la vie," *Bulletin de l'Académie de médecine*, 2nd ser., 7 (1878): 626.

57. For the British reception of Marey's apparatus, see Christopher Lawrence, "Incommunicable Knowledge: Science, Technology and the Clinical Art in Britain 1850–1914," *Journal of Contemporary History* 20 (1985): 503–20. For the French clinicians, see the next chapter.

58. Félix Giraud-Teulon, "Mécanique animale: Étude retrospective sur les progrès realisés dans l'histoire des mécanismes de la locomotion chez l'homme, par la méthode des inscriptions graphiques de M. le professeur Marey," *Bulletin de l'Académie de médecine*, September 4, 1883, 1029.

59. See Victor Legros, *Étude expérimentale de la marche: Section I; Marche d'un homme isolé* (Paris: Tanera, 1876), who praised Marey's apparatus but found the simple podometer to be a "more modest and maniable instrument" (24).

60. Marey, *La machine animale*, 153.

61. Marey, *Animal Mechanism*, 152, 163. See also the later series of images that Marey showed at his lecture in 1878, which apparently derived from Lenoble Du Teil.

62. The point has been made, correctly, that the graphical method, like other forms of notation, also necessarily relied on conventions. See Chadarevian, "Graphical Method and Discipline," 44–45. Nevertheless, Marey boasted that his method was superior to all other systems "because, so to speak, it does not borrow anything from convention." (Marey, *La méthode graphique*, 1–2).

63. See Raabe's unpublished letters to Marey (February 8, 11, and 13, 1877, Musée Marey, Beaune). Marey's responses have not survived, but some of their content can be reconstructed. In 1882 Raabe supplied the Station with a small horse, of which several series of chronophotographs were produced.

64. Jules Lenoble Du Teil, *Locomotion quadrupède étudiée sur le cheval* (Paris: Bureau du Journal des Haras, 1877), i.

65. In *La méthode graphique*, Marey discussed Ibry's diagrams, which showed the travel of all trains of one line according to the hours of the day in synoptic fashion. See Marey, *La méthode graphique*, 21, and Léon Lalanne, "Méthodes graphiques: Note sur un nouveau mode de représentation de la marche des trains sur une voie de communication," *Comptes rendus des séances de l'Académie des sciences* 99 (1884): 307–13.

66. See Marey's critique in *La méthode graphique*, 438.

67. Barroil, *L'art équestre: Première partie*, 90–91. The correspondence between Bonnal und Demenÿ shows that Bonnal criticized the Station physiologique apparatus as early as 1882. See the letters of July 25 and August 3, 1882, in Lefebvre, Malthête, and Mannoni, *Lettres d'Etienne-Jules Marey à Georges Demenÿ*, 459–61.

68. Henri Bonnal, "Préface," in Barroil, *L'art équestre: Deuxième partie* xviii.

69. Jules Lenoble Du Teil, *Examen de l'art equestre de M. E. Barroil (Doctrine du Capitaine Raabe): Avec rectification de ses théories, d'après la méthode expérimentale* (Paris: Bureaux de la Revue des Haras, 1890), 65.

70. See chap. 3, pp. 87–88.

71. Richard Volkmann, "Die Krankheiten der Bewegungsorgane," in *Handbuch der allgemeinen und speciellen Chirurgie*, ed. Franz von Pitha and Theodor Billroth (Erlangen: Enke, 1872), 2:722–27. Duchenne, *Physiology of Motion*, chap. 7 (303–439).

72. Central to these reforms was the work of Hermann Meyer (see chap. 3) and his "rational shoes" conceived for the Swiss army: *Die richtige Gestalt der Schuhe: Eine Abhandlung aus der angewandten Anatomie für Aerzte und Laien geschrieben* (Zurich: Meyer & Zeller, 1858); Samuel Auguste Salquin, *Die militärische Fussbekleidung* (Bern: Jent & Reinert, 1881); Charles Viry, "De la chaussure du soldat d'infanterie," *Archives de médecine et de pharmacie militaires* 9 (1887): 1–18.

73. Ernest Onimus, "Des déformations de la plante des pieds, spécialement chez les enfants, dans les affections atrophiques et paralytiques de la jambe: Mémoire lu à l'Association française pour l'avancement des sciences, dans la séance du 19 août 1876," *Gazette hébdomadaire de médecine et de chirurgie* 34 (1876): 532.

74. Ibid., 531.

75. Ibid., 532.

76. Ernest Onimus, "Étude physiologique et clinique des surfaces en contact avec le sol," *Revue de médecine* 1 (1881): 658–59. Onimus also extended his clinical material by drawing on ethnographic observations of the Kabyle people made by Duhousset in Algeria, seeking to identify differences of gait based on race. Ernest Onimus, "Des déformations du pied et de la jambe," *Revue de chirurgie* 2, no. 6 (1882): 443–62; 652–72. Emile Duhousset, "Études sur les Kabyles du Djurjura," *Bulletins de la Société d'anthropologie de Paris* 2, no. 3 (1868): 265–71.

77. Onimus, "Des déformations de la plante des pieds," 533.

78. Ernest Onimus and Charles Viry, *Étude critique des tracés obtenus avec le cardiographe et le sphygmographe* (Paris: Germer Baillière, 1866).

79. Joseph Rohmer, *Les variations de forme normales et pathologiques de la plante du pied étudiées par la méthode graphique* (Nancy: Collin, 1879). In David Hayes Agnew, *The Principles and practice of surgery* (Philadelphia: J. B. Lippincott, 1889), 3:363, David Hayes Agnew lists Marey, Ludwig, Duhousset, Onimus, and Rohmer as pioneers of the "graphic method, as it is termed, or the method of studying the physiological and pathological action of the muscles by taking impressions of the feet."

80. Jean-Martin Charcot, *Lectures on the Diseases of the Nervous System*, trans. George Sigerson (London: New Sydenham Society, 1881), 4.

81. The first textbooks that provided classical descriptions of disturbances in motor coordination and gait were Moritz Heinrich Romberg, *Lehrbuch der Nerven-krankheiten des Menschen*, 2 vols. (Berlin: Duncker, 1840); published in English as *A Manual of the Nervous Diseases of Man*, trans. Edward H. Sieveking (London: Sydenham Society, 1853); and Sir William Gowers *A Manual of Diseases of the Nervous System*, 2 vols. (London: Churchill, 1886–1888). See Stephen T. Casper, "The Patient's Pitch: The Neurologist, the Tuning Fork, and Textbook Knowledge," in *The Neurological Patient in History*, ed. L. Stephen Jacyna and Stephen T. Casper (Rochester, NY: University of Rochester Press, 2012), 21–43.

82. See Christopher G. Goetz, Michel Bonduelle, and Toby Gelfand, *Charcot: Constructing Neurology* (Oxford: Oxford University Press, 1995), 78–79; Jean-Martin Charcot, *Leçons du mardi à la Salpêtrière: Policlinique 1887–1888*, 2nd ed., vol. 1 (Paris: Bureaux du progrès medical, 1892). For a partial English translation, see Christopher G. Goetz, *Charcot, the Clinician: The Tuesday Lessons; Excerpts from Nine Case Presentations on General Neurology Delivered at the Salpêtrière Hospital in 1887–88 by Jean-Martin Charcot* (New York: Raven Press, 1987).

83. Charcot, *Leçons du mardi*, 170. However, one should not forget that the broad reception of Charcot's teaching and his dramatic rhetorical style (to which Freud added considerably in his German translations), wherein patient presentations are conveyed through dialogic exchange with the physician, was ultimately channeled through the publication of a book. The written version of Charcot's lectures was based on transcriptions by his pupils and was later revised and corrected by Charcot himself for a second printing.

84. In that sense, it seems misleading to label the different contributions of Charcot and his school with this general term, as is often done in the literature.

85. Jean-Martin Charcot, *Clinical Lectures on Certain Diseases of the Nervous System*, trans. Edward P. Hurd (Detroit: G. S. Davis, 1888), 32–48.

86. See especially Paul Richer, *Études sur l'attaque hystéro-épleptique faites à l'aide de la méthode graphique* (Paris: Delahaye, 1879), who used Marey's self-recording instruments in collaboration with the physicist Paul Regnard. Their use was later mostly restricted to the control of simulation in the hypnosis experiments. See Andreas Mayer, *Sites of the Unconscious: Hypnosis and the Emergence of the Psychoanalytic Setting*, trans. Christopher Barber (Chicago: University of Chicago Press, 2013), 37–46.

87. Charcot, *Leçons du mardi*, 3–4; English translation in Goetz, *Charcot, the Clinician*, 5. At another point, Charcot notes that "stepper" in English designates a "cheval qui a de l'action," that is, a horse trained to lift its feet high off the ground while walking or trotting (*Leçons du mardi*, 203).

88. Charcot, *Leçons du mardi*, 14.

89. See, for example, the discussion of Friedreich's ataxia, a degenerative illness of the central nervous system whose multiplicity of symptoms made it difficult to detect, named after the German physician Nikolaus Friedreich, who first described it in 1862. Charcot, *Leçons du mardi*, 174–83; Goetz, *Charcot, the Clinician*, 141–53.

90. Charcot, *Leçons du mardi*, 204; Goetz, *Charcot, the Clinician*, 5n7.

91. On the role of photography at the Salpêtrière, see the perceptive essay by André Gunthert, "Klinik des Sehens: Albert Londe, Wegbereiter der medizinischen Fotografie," *Fotogeschichte* 21 (2001): 27–40, and Denis Bernard and André Gunthert, *L'instant rêvé, Albert Londe* (Nîmes: Jacqueline Chambon, 1993).

92. Georges Gilles de la Tourette, *Études cliniques et physiologiques sur la marche: La marche dans les maladies du système nerveux étudiée par la méthode des empreintes; Thèse pour le doctorat en médecine présentée et soutenue le lundi 28 décembre 1885 à 2 heures* (Paris: Imprimerie de la Société de typographie, 1885). Besides Onimus' studies, the major reference for Gilles de la Tourette's new method was Franz Neugebauer, *Zur Entwickelungsgeschichte des spondylolisthetischen Beckens und seiner Diagnose (mit Berücksichtigung von Körperhaltung und Gangspur)*, (Dorpat: Niemeyer, 1881); "Zur Casuistik des sogenannten spondylolisthetischen Beckens," *Archiv für Gynäkologie* 19, no. 3 (1882): 441–74; and "Aetiologie der sogenannten Spondylolisthesis," *Archiv für Gynäkologie* 20, no. 3 (1883): 133–84. On Gilles de la Tourette, see A. Lees: "Georges Gilles de la Tourette. The Man and His Times," *Revue neurologique* 142, no. 11 (1986): 808–16.

93. Gilles de la Tourette, *Études cliniques*, 18. It has to be noted, however, that the clinical applications were closely circumscribed with the aim of rendering a clear, visible expression of the clinical types. Because of the high variability of the movements involved in chorea, Gilles de la Tourette excluded patients suffering from this condition from his study (13). In the following years, he would nevertheless extend his method to cases of hysteria. See Georges Gilles de la Tourette, "L'attitude de la marche dans l'hémiplégie hystérique," *Nouvelle iconographie de la Salpêtrière* 1 (1888): 1–12.

94. Hermann Vierordt, *Das Gehen des Menschen in gesunden und kranken Zuständen nach selbstregistrierenden Methoden dargestellt* (Tübingen: H. Laupp'sche Buchhandlung, 1881).

95. Ibid, 10.

96. Gilles de la Tourette, *Études cliniques*, 22.

97. Étienne-Jules Marey, "Des forces utiles dans la locomotion (Conférence faite au congrès international d'hygiène de La Haye, septembre 1884," *Revue scientifique* 34 (1884): 515–16.

98. See the letter by Marey to Demenÿ, January 2, 1886, in Jacques Malthête, and Laurent Mannoni, *Lettres d'Etienne-Jules Marey à Georges Demenÿ*, 169–70.

99. Georges Demenÿ and Edouard Quénu, "Étude de la locomotion humaine dans les cas pathologiques," *Comptes rendus des séances de l'Académie des sciences* 107 (1888): 1550–64; "De la locomotion dans l'ataxie locomotrice," *Comptes rendus des séances de l'Académie des sciences* 108 (1889), 963–64; Louis Gastine: *La chronophoto-graphie sur plaque fixe et sur pellicule mobile* (Paris: Gauthier-Villars / Masson, 1897), 73–83.

100. Thus Cesare Lombroso, in his popular criminal anthropology, tried to prove that the typical criminal possessed characteristic deformities of the toes reminiscent of the "prehensile feet" of apes and monkeys, an "atavistic anomaly, " according to his theories. Cesare Lombroso, *Criminal Man*, trans. Mary Gibson and Nicole Hahn Rafter (Durham, NC: Duke University Press, 2006), 308. Nevertheless, one should note that the footprint in forensics is concerned with the individual criminal and not with a criminal type; see Hans Gross's *Handbuch der Kriminalistik*, 8th ed. (Berlin: J. Schweitzer, 1908), 582–650, where several methods of preserving or reconstructing footprints as evidence are discussed at length.

101. Paul Richer, *Introduction à la figure humaine* (Paris: Gaultier / Magnier, 1902), 111.

102. Ibid.

103. Paul Richer: *Physiologie artistique de l'homme en movement* (Paris: Doin, 1895).

104. Marey, "Moteurs animés," 294; *Die Chronophotographie*, trans. Adolf von Heydebreck (Berlin: Mayer & Müller, 1893), 56. Marey also corresponded on this subject with the famous Hungarian academic painter Bertalan Székely. See Alexandre Métraux, "Das eidetische Pferd: Wie Bertalan Székely Bildanimationen konstruierte," in *Kunst-maschinen: Spielräume des Sehens zwischen Wissenschaft und Ästhetik*, ed. Andreas Mayer and Alexandre Métraux (Frankfurt: Fischer, 2005), 61–100.

105. See Michel Poivert, "Variété et vérité du corps humain: L'esthétique de Paul Richer," in *L'art du nu au XIXe siècle: Le photographe et son modèle* (Paris: Hazan, 1997), 164–75. On the culture of male athletes in France and their influence on art edu-cation, see Tamar Garb, *Bodies of Modernity: Figure and Flesh in Fin-de-Siècle France* (London: Thames & Hudson, 1998) 55–79.

106. Duhousset, "Proportions comparatives de l'homme et du cheval," *Gazette des Beaux-Arts* 5, no. 3 (1891): 385–400; Paul Richer, *Canon des proportions du corps humain* (Paris: Delagrave, 1893).

107. Paul Richer, *Introduction à la figure humaine*, 130–31.

108. Marey, *La méthode graphique*, iv. See also the critical discussion of representa-tions of horse movement in ancient art in Marey's lecture "Moteurs animés."

109. Salomon Reinach, "Chronique d'Orient," *Revue archéologique* 9 (1887): 107, with reference to Ernst Curtius's thesis *Die knieenden Figuren der altgriechischen*

Kunst (Berlin: Archäologische Gesellschaft, 1869). In a later important study, Reinach addresses galloping horse images in ancient and modern art at length: Salomon Reinach, "La représentation du galop dans l'art ancien et modern," *Revue archéologique* 36 (1900): 216–51, 441–50; 37 (1900): 244–59; 38 (1901): 27–45; 39 (1901): 9–11. By around 1900, the archaeologist occupied by "pedestrian researches" was already a subject of literary parody, as becomes evident when one reads Wilhelm Jensen's novella *Gradiva* (1903) not as a case of foot fetishism (as in Freud's famous later interpretation) but rather in the context of its mixed reception by classical scholars of the period. See Andreas Mayer, "Gradiva's Gait: Tracing the Figure of a Walking Woman," *Critical Inquiry* 38, no. 3 (2012): 554–78.

110. Maurice Emmanuel, *Essai sur l'orchestique grecque: Étude de ses mouvements d'après les monuments figurés* (Paris: Hachette, 1895).

111. Translator's note in Marey, *Die chronophotographie*, 56.

112. Ibid., 55–56. Von Heydebreck had dealt with this subject in an extended essay, "Über die Grenzen von Malerei und Plastik," *Verhandlungen der philosophischen Gesellschaft zu Berlin* 7/8 (1878): 1–41.

113. Translator's note in Marey, *Die chronophotographie*, 61.

114. Ernst Brücke, *Schönheit und Fehler der menschlichen Gestalt*, 2nd ed. (Vienna: Braumüller, 1893), 2–3.

115. See Emil du Bois-Reymond, "Naturwissenschaft und bildende Kunst" (1890), in *Reden von Bois-Reymond*, ed. Estelle Bois-Reymond (Leipzig: Veit, 1912), 2:407–09.

116. Ernst Brücke, "Die Darstellung der Bewegung durch die bildenden Künste," *Deutsche Rundschau* 26 (1881): 46–47.

117. Ibid., 54.

118. In the case of Brücke's colleague Exner, for example, the register of *subjective Gesichtserscheinungen* (subjective visual perceptions). Sigmund Exner, *Physiologisches und Pathologisches in den bildenden Künsten*, (Vienna: Verein zur Verbreitung naturwissenschaftlicher Kenntnisse, 1889), 8. By contrast, the art historian Alois Riegl answered Paul Richer's "objective aesthetics" with a psychological interpretation of various epochs in art history; see "Objektive Ästhetik," *Literatur-Blatt der Neuen Freien Presse*, July 13, 1902.

119. Claude Blaenckart, ed., *Les politiques de l'anthropologie: Discours et pratiques en France 1860–1940* (Paris: L'Harmattan, 2001); Nélia Dias, *La mesure des sens: Les anthropologues et le corps humain au XIXe siècle* (Paris: Aubier / Flammarion, 2004).

120. Duhousset, "Études sur les Kabyles," 269.

121. On Manouvrier, see Jennifer Hecht, *The End of the Soul: Scientific Modernity, Atheism, and Anthropology in France* (New York: Columbia University Press, 2005), 211–56. Some of Regnault's work is discussed in Fatimah Tobing Rony, *The Third Eye: Race, Cinema, and Ethnographic Spectacle* (Durham, NC: Duke University Press, 1996), but it is largely reduced to the function of reproducing the asymmetry between the European observers and the "savages," serving the ideological representation of the "Other."

122. Léonce Manouvrier, "Étude sur la rétroversion de la tête du tibia et l'attitude humaine à l'époque quaternaire," *Mémoires de la Société d'anthropologie de Paris*

4 (1893): 219–64; "La platycnémie chez l'homme et chez les singes," *Bulletins de la Société d'anthropologie de Paris* 10 (1887): 128–41.

123. Félix Regnault, "La marche et le pas gymnastique militaires," *La Nature,* July 29, 1893, 129.

124. Félix Regnault, "Exposition ethnographique de l'Afrique occidentale au Champs-de-Mars à Paris: Sénégal et Soudan français," *La Nature,* August 17, 1895, 183–86; "Le grimper," *Revue encyclopédique* (October 23, 1897): 904–5.

125. Charles Comte and Félix Regnault, "Marche et course en flexion," *Bulletins de la Société d'anthropologie de Paris* 7, no. 7 (1896): 337.

126. Charles Comte und Felix Regnault, "Étude comparative entre la méthode de marche et de course dite de fléxion et les allures ordinaires," *Archives de physiologie* (1896): 380–89.

127. See his manual of running: Louis-Firmin Lafontaine [Firmin Weiss], *Manuel-théorie des courses à pied* (Sens: Modrine, 1888).

128. See Napoleon Laisné, *Gymnastique pratique, contenant la description des exercices, la construction et le prix des machines, et des chants spéciaux inédits: Ouvrage destiné aux familles, aux établissements d'éducation, aux corps militaires* (Paris: Dumaine, 1850).

129. Félix Regnault, "Des différentes manières de marcher," *Bulletins de la Société d'anthropologie de Paris* 4 (1893): 382.

130. Félix Regnault and Albert-Charlemagne-Oscar de Raoul, *Comment on marche: Des divers modes de progression; De la superiorité du mode en flexion* (Paris: Lavauzelle, 1898), 22–23; Félix Regnault, "Du pas gymnastique," *La Nature ,* January 6, 1894, 83–86; January 20, 1894, 122–23.

131. Regnault and Raoul, *Comment on marche,* 55. In this context, it is significant that Regnault reprinted in an appendix the chapter "Suggestion in war" (ibid., 141–66) from his earlier book *Hypnotisme, Religion* (Paris: Reinwald, 1897).

132. See Weisbach, "Die sogenannte 'Fussgeschwulst'—Syndesmitis metatarsea—des Infanteristen in Folge von anstrengenden Märschen," *Deutsche militär-ärztliche Zeitschrift* 6 (1877): 551–57; Rittershausen, "Die Marschgeschwulst oder das sogenannte Ödem des Mittelfusses," *Militär-Wochenblatt* 75 (1894); Martin Kirchner, *Grundriss der Militärgesundheitspflege* (Braunschweig: Bruhn, 1896), 1142–44; Nathan Zuntz and Wilhelm August Ernst Schumburg, *Studien zu einer Physiologie des Marsches* (Berlin: August Hirschwald, 1901); Franz Thalwitzer, *Der Parademarsch: Eine ärztliche Betrachtung* (Dresden: Paul Alicke, 1904). Some physicians pointed out that the characteristics of the goose step—the effort to keep the head and trunk completely immobile and to march in line—must lead not only to physical exhaustion of the soldiers but also to heightened "psychical strain." Kirchner, *Grundriss der Militärgesundheitspflege,* 1132.

133. Regnault and de Raoul, *Comment on marche,* 23.

134. Étienne-Jules Marey, "Préface," in Regnault and de Raoul, *Comment on marche,* 6.

135. Ibid., 7–8.

136. Gustave Le Bon, an early enthusiast of Marey's graphical method and a zeal-ous advocate of Baucher's methods of dressage, later used chronophotography in his own studies of the horse in which he also sketched a "psychology of obedience" of the

horse couched in the language of associationist psychology. See *L'equitation actuelle et ses principes: Recherches expérimentales* (Paris: Firmin-Didot, 1892). In this treatise, an atlas of chronophotographies produced by Albert Londe served to demonstrate the transformation achieved by Le Bon's method of rectifying the horse's position in motion. One of the key figures to use hypnotic suggestion in children's education was Edgar Bérillon (1859–1948), founder of the Institut psycho-physiologique in Paris in 1892. See his *De la suggestion et de ses applications à la pédagogie* (Paris: Bureaux de la Revue de l'Hypotisme, 1888).

137. During the following years, the *marche en flexion* was promoted by Regnault in a private sanatorium as a new form of therapy that could cure all sorts of evils brought about by modern urban life (Félix Regnault, "Méthode de la course en flexion (Dromothérapie)," *Gazette médicale de Paris*, 12th ser., 3, no. 43 (October 24, 1903): 349–50. According to Paul Richer, who criticized the two reformers for overlooking the complexities of muscular contraction and presenting an oversimplified mechanical view of labour and fatigue, their method was abandoned after later experiments in the army on the recommendation of the famous military physician Alphonse Laveran. Paul Richer, *Nouvelle anatomie artistique III: Cours supérieur (suite); Physiologie; Attitude et mouvements* (Paris: Plon, 1921), 157.

138. Felix Regnault, "Le langage par le geste," *La Nature*, October 15, 1898, 315–17; "La chronophotographie dans l'ethnographie," *Bulletins et mémoires de la Société d'anthropologie de Paris*, 5th ser., 1 (1900): 422.

CONCLUSION

1. Sigfried Giedion, *Mechanization Takes Command: A Contribution to Anonymous History* (New York: Oxford University Press, 1948).

2. After Braune's death in 1892, Otto Fischer would sign many of the later contributions with his own name. See Wilhelm Braune and Otto Fischer, "Der Gang des Menschen I. Theil: Versuche am unbelasteten und belasteten Menschen," *Abhandlungen der mathematisch-physischen Classe der königlich sächsischen Gesellschaft der Wissenschaften* 21, no. 4 (1895): 153–322, and the other studies by Fischer listed in the bibliography.

3. "We may assume that Marey's first experimental subject also walked during the chronophotographic recording in a gait that corresponds to the brisk progression adopted on a country road." Otto Fischer, "Der Gang des Menschen V. Theil: Die Kinematik des Beinschwingens," *Abhandlungen der mathematisch-physischen Classe der königlich sächsischen Gesellschaft der Wissenschaften* 28, no. 5 (1903): 334.

4. Otto Fischer, "Der Gang des Menschen VI. Theil: Über den Einflusz der Schwere und der Muskeln auf die Schwingungsbewegung des Beins," *Abhandlungen der mathematisch-physischen Classe der königlich sächsischen Gesellschaft der Wissenschaften* 28, no. 7 (1904): 616.

5. Andreas Mayer, *Sites of the Unconscious: Hypnosis and the Emergence of the Psychoanalytic Setting*, trans. Christopher Barber (Chicago: University of Chicago Press, 2013). For trauma research, see Ruth Leys, *Trauma: A Genealogy* (Chicago: University of Chicago Press, 1999).

6. Josef Breuer and Sigmund Freud, *1893–1895: Studies on Hysteria*, vol. 2 of *The Standard Edition of the Complete Psychological Works of Sigmund Freud*, trans. and ed. James Strachey (London: Hogarth Press / Institute of Psycho-Analysis, 1955), 135.

7. Ibid., 148.

8. Ibid., 179.

9. Italo Svevo: *Zenos Gewissen*, trans. Barbara Kleiner (Frankfurt: Zweitausendeins, 2000), 144.

10. Ibid.

11. George Humphrey, *The Story of Man's Mind* (Boston: Small, Maynard, 1923), 109.

12. Gustav Meyrink, "Der Fluch der Kröte – Fluch der Kröte," in *Gesammelte Werke*, vol. 4, pt. 2, (Munich: Langen, 1913), 216–21.

13. E. Ray Lankester, "The Muybridge Photographs," *Nature* 40 (May 23, 1889): 78–80.

14. Wilhelm Steinhausen, "Mechanik des menschlichen Körpers (Ruhelagen, Gehen, Laufen, Springen)," in *Bewegung und Gleichgewicht, Physiologie der körperlichen Arbeit I: Handbuch der normalen und pathologischen Physiologie*, ed. A. Bethe et al., vol. 15, pt. 1 (Berlin: Springer, 1930), 164.

15. For a general account of this trend in German science (with no reference to locomotion science), see Anne Harrington, *Reenchanted Science: Holism in German Culture from Wilhelm II to Hitler* (Princeton, NJ: Princeton University Press, 1996). A holistic approach to the *Bewegungsgestalt* is developed by the Dutch philosopher and anthropologist Frederik J. J. Buytendijk in *Allgemeine Theorie der menschlichen Handlung und Bewegung als Verbindung und Gegenüberstellung von physiologischer und psychologischer Betrachtungsweise* (Berlin, New York: Springer, 1972 [1948]).

16. Marcel Mauss, "Techniques of the Body," 75.

17. One may note that his commentators—beginning with Claude Lévi-Strauss in his famous *Introduction to the Work of Marcel Mauss*, trans. Felicity Baker (London: Routledge & Kegan Paul, 1987) [1950])—have also completely ignored the context of the science of locomotion. The most recent (and excellent) introduction by Nathan Schlanger to Marcel Mauss, *Techniques, Technology, and Civilization* (New York: Berghahn Books, 2006), 1–29, contextualizes the essay exclusively with regard to the human and social sciences.

18. Mauss, "Techniques of the Body," 76.

19. Ibid., 73 (translation modified).

20. See above, pp. 137–41.

21. Mauss, "Techniques of the Body," 72.

22. See especially Marcel Mauss, *Manuel d'ethnographie* (Paris: Payot, 1967 [1947]), 30, a manual compiled from students' notes by Denise Paulme. In the new edition by Nathan Schlanger, which incorporates some unpublished drafts by Mauss, this feature becomes even more apparent but does not receive any closer analysis. "One has to draw oneself [*dessiner soi-même*]," insists Mauss in one his lectures, "for the photo does not replace a good sketch, just as the cinema does not replace the successive outlines [*schémas successifs*] of a movement." Marcel Mauss, "Cours d'ethnographie descriptive," quoted in Mauss, *Techniques, technologie et civilisation*, 14.

BIBLIOGRAPHY

Adamson, James. *Sketches of Our Information as to Railroads*. Newcastle: Edward Walker, 1826.

Adelon, Nicolas Philibert. *Physiologie de l'homme*. 4 vols. Paris: Compère jeune, 1823.

Agnew, David Hayes. *The Principles and Practice of Surgery*. 3 vols. Philadelphia: J. B. Lippincott, 1889.

Albinus, Bernhard Siegfried. *Tabulae sceleti et musculorum corporis humani*. Lugduni Batavorum: Johannem & Hermannum Verbeek, 1747.

Ambrière, Madeleine. *Au soleil du romantisme: Quelques voyageurs de l'infini*. Paris: Presses universitaires de France, 1998.

Ambrière, Madeleine. *Balzac et la recherche de l'absolu*. Paris: Presses universitaires de France, 1999.

Ambrière, Madeleine. "Balzac le chercheur d'absolu: Du coté de la science." In Ambrière, *Au soleil du romantisme*, 213–306.

Ambrière, Madeleine. "Les enchanteurs de la science: Balzac et les Messieurs du Muséum (1833)." In Ambrière, *Au soleil du romantisme*, 230–48.

Anon. "Antikritik gegen die in der Allg. Deutsch. Bibl. befindliche Recension, über die Reine Taktik des Rittm. und Flgladj. v. Miller." *Intelligenzblatt der Allgemeinen Literatur-Zeitung* , May 19, 1790, col. 506–12.

Anon. "Kriegswissenschaften. Stuttgart, in der Druckerey der hohen Karlsschule: Reine Taktik der Infanterie, Cavallerie und Artillerie, in zwey Theilen verfasst von Franz Miller." *Allgemeine Literatur-Zeitung*, March 8, 1789, 577–79.

Anon. "Über Chausseedampfwagen und Pferdeeisenbahnen." *Polytechnisches Journal* 52 (1834): 401–8.

Anon. "Über Dampfwagen." *Bayerische National-Zeitung*, October 4, 1834, 1092–94; October 5, 1834, 1096–97.

Anon. "Ueber selbstfahrende Fuhrwerke." *Polytechnisches Journal* 55 (1835): 12–16.

Ariès, Philippe. *Centuries of Childhood: A Social History of Family Life*. Translated by Robert Baldick. New York: Alfred A. Knopf, 1962.

"Barthez." In *Nouvelle biographie des contemporains, ou Dictionnaire historique et raisonné de tous les hommes qui, depuis la Révolution Française, ont acquis de la célébrité par leurs actions, leurs écrits, leurs erreurs ou leurs crimes, soit en France, soit dans les pays étrangers,* edited by Arnault, Antoine-Vincent, Antoine Jay, Étienne de Jouy, and Jacques Marquet de Norvins, vol. 2. Paris: Libraire Historique, 1820.

[Aumont, Arnulphe d']. "Debout." In *Encyclopédie ou Dictionnaire raisonné des sciences, des arts et des métiers.* Vol. 4. Edited by Denis Diderot and Jean-Baptiste le Rond d'Alembert, 654–57. Paris: Briasson, 1754.

Azouvi, François. *Maine de Biran: La Science de l'homme.* Paris: Vrin, 1995.

Baader, Joseph Ritter von. *Die Unmöglichkeit, Dampfwagen auf gewöhnlichen Straßen mit Vortheil als allgemeines Transportmittel einzuführen, und die Ungereimtheit aller Projekte, die Eisenbahnen dadurch entbehrlich zu machen.* Nuremberg: Riegel und Wießner, 1835.

Balzac, Honoré de. *Abglanz meines Begehrens: Bericht einer Reise nach Russland 1847.* Berlin: Friedenauer Presse, 2018.

Balzac, Honoré de. *Correspondance I (1809–1835).* Paris: Gallimard, 2006.

Balzac, Honoré de. *Ferragus.* In *Œuvres completes,* vol. 5, 793–904. Paris: Gallimard, 1977.

Balzac, Honoré de. *Ferragus, Chief of the Dévorants.* Translated by Katharine Prescott Wormeley. Project Gutenberg, 2004.

Balzac, Honoré de. *La peau de chagrin.* In *La comédie humaine,* vol 10, 3–294. Paris: Gallimard, 1979.

Balzac, Honoré de. *Physiologie du mariage.* In *Œuvres completes,* vol. 11, 865–1205. Paris: Gallimard, 1980.

Balzac, Honoré de. *Théorie de la démarche.* In *La comédie humaine,* vol. 12, 259–302. Paris: Gallimard, 1981.

Balzac, Honoré de. *Traité de la vie élégante."* In *La comédie humaine,* vol. 12, 211–57. Paris: Gallimard, 1981.

Barroil, Étienne. *L'art équestre: Première partie; Iconographie des allures et des changements d'allures.* Paris: J. Rothschild, 1889.

Barroil, Étienne. *L'art équestre: Deuxième partie; Dressage raisonné du cheval; avec une préface du Commandant Bonnal.* Paris: J. Rothschild, 1889.

Barthez, Paul-Joseph. "Éclaircissemens sur quelques points de la mécanique des mouvemens de l'homme." *Mémoires de la Société médicale d'emulation* 11 (1803): 259–80.

Barthez, Paul-Joseph. *Nouveaux eléments de la Science de l'homme.* 3rd ed. 2 vols. Paris: Germer Ballière, 1858 [1778].

Barthez, Paul-Joseph. *Nouvelle mécanique des mouvements de l'homme et des animaux.* Carcassonne: de l'Imprimerie de Pierre Polère, 1798.

Bastholm, Eyvind. *History of Muscle Physiology.* Copenhagen: Munksgaard, 1950.

Baucher, François. *Dictionnaire raisonné de l'équitation.* 2nd ed. Paris: F. Baucher, 1851 [1833].

Bausinger, Hermann, Klaus Beyrer, and Gottfried Korff, eds. *Reisekultur: Von der Pilgerfahrt zum modernen Tourismus.* Munich: C. H. Beck, 1991.

Beckmann, Johann. *Beyträge zur Geschichte der Erfindungen.* Leipzig: Paul Gotthelf Kummer, 1782.

Benjamin, Walter. *The Arcades Project*. Translated by Howard Eiland. Cambridge, MA: Harvard University Press, 2002.

Benjamin, Walter. *Das Passagen-Werk*. Vol. 5 of *Gesammelte Schriften*, edited by Rolf Tiedemann. Frankfurt: Suhrkamp, 1982.

Bérillon, Edgar. *De la suggestion et de ses applications à la pédagogie*. Paris: Bureaux de la Revue de l'Hypotisme, 1888.

Bernard, Denis, and André Gunthert. *L'instant rêvé, Albert Londe*. Nîmes: Jacqueline Chambon, 1993.

Bernett, Hajo. *Die pädagogische Neugestaltung der bürgerlichen Leibesübungen durch die Philanthropen*. 3rd. ed. Schorndorf bei Stuttgart: Hoffmann, 1971.

Bernoulli, Christoph. *Elementarisches Handbuch der industriellen Physik, Mechanik und Hydraulik*. 2 vols. Stuttgart: Cottasche Buchhandlung, 1835.

Bessel, Friedrich Wilhelm. *Populäre Vorlesungen über wissenschaftliche Gegenstände*. Hamburg: Perthes-Besser & Mauke, 1848.

Beyrer Klaus. *Die Postkutschenreise*. Tübingen: Tübinger Vereinigung für Volkskunde, 1985.

Bichat, Xavier. *Anatomie descriptive*. Rev. ed. 5 vols. Paris: Gabon, 1829.

Bichat, Xavier. *Recherches physiologiques sur la vie et la mort*. Paris: Brosson, Gabon, 1800.

Blaenckart, Claude ed. *Les politiques de l'anthropologie: Discours et pratiques en France 1860–1940*. Paris: L'Harmattan, 2001.

Blumenberg, Hans. *Die Lesbarkeit der Welt*. 2nd ed. Frankfurt: Suhrkamp, 1983.

Bois-Reymond, Emil du. "Naturwissenschaft und bildende Kunst" (1890). In vol. 2 of *Reden von Bois-Reymond*, edited by Estelle Bois-Reymond, 407–9. Leipzig: Veit, 1912.

Boissier de Sauvages, François. *Nosologia methodica sistens morborum classes, genera et species, juxta Sydenhami mentem et Botanicorum ordinem*. Amsterdam: Frères De Tournes, 1763.

Boissier de Sauvages, François. *Nosologie méthodique, dans laquelle les maladies sont rangées par classes, suivant le système de Sydenham, & l'ordre des botanists*. 10 vols. Paris: Hérissant le fils, 1771.

Borelli, Giovanni Alfonso. *De motu animalium*. Rome: Angelo Bernabò, 1680–1681.

Borelli, Giovanni Alfonso. *De motu animalium: Editio nova, a plurimis mendis repurgata*. The Hague: Petrum Gosse, 1743.

Borelli, Giovanni Alfonso. *On the Movement of Animals*. Translated by Paul Maquet. New York: Springer, 1989.

Borscheid, Peter. *Das Tempo-Virus: Eine Kulturgeschichte der Beschleunigung*. Frankfurt: Campus, 2004.

Böttger, Heinrich Ludwig Christian. "Vorschlag einer Uniform für Reisende zu Fuße." *Journal des Luxus und der Moden* 15 (May 1800): 217–23.

Brain, Robert. "The Graphic Method: Inscription, Visualization, and Measurement in Nineteenth-Century Science and Culture." PhD diss., University of California, Los Angeles, 1996.

Braun, Marta. *Picturing Time: The Work of Etienne-Jules Marey (1830–1904)*. Chicago: University of Chicago Press, 1992.

Braun, Rudolf, and David Gugerli. *Macht des Tanzes, Tanz der Mächtigen: Hoffeste und Herrschaftszeremoniell 1550–1914*. Munich: C. H. Beck, 1993.

Braune, Wilhelm, and Otto Fischer. "Der Gang des Menschen I. Theil: Versuche am unbelasteten und belasteten Menschen." *Abhandlungen der mathematisch-physischen Classe der königlich sächsischen Gesellschaft der Wissenschaften* 21, no. 4 (1895): 153–322.

Bremmer, Jan. "Walking, Standing, and Sitting in Ancient Greek Culture." In *A Cultural History of Gesture from Antiquity to the Present Day*. Edited by Jan Bremmer and Herman Roodenburg, 15–35. Ithaca, NY: Cornell University Press, 1991.

Breuer, Josef, and Sigmund Freud. *1893–1895: Studies on Hysteria*. Vol. 2 of *The Standard Edition of the Complete Psychological Works of Sigmund Freud*, translated and edited by James Strachey. London: Hogarth Press / Institute of Psycho-Analysis, 1955.

Brillat-Savarin, Jean Anthèlme. *Physiologie du goût, ou Méditations de gastronomie transcendante: Ouvrage théorique, historique et à l'ordre du jour, dédié aux gastronomes parisiens par un professeur, membre de plusieurs sociétés savantes*. 3rd ed. 2 vols. Paris: A. Sautelet, 1829.

Broca, Paul. *Eloge historique de P. N. Gerdy*. Paris: Bureau du Moniteur des Hôpitaux, 1856.

Brown, Theodore M. *The Mechanical Philosophy and the "Animal Oeconomy."* New York: Arno, 1981.

Brücke, Ernst. "Die Darstellung der Bewegung durch die bildenden Künste." *Deutsche Rundschau* 26 (1881): 39–54.

Brücke, Ernst. *Schönheit und Fehler der menschlichen Gestalt*. 2nd ed. Vienna: Braumüller, 1893.

Brune, Thomas. "Von Nützlichkeit und Pünktlichkeit der Ordinari-Post." In *Reisekultur: Von der Pilgerfahrt zum modernen Tourismus*, edited by Hermann Bausinger, Klaus Beyrer, and Gottfried Korff, 123–30. Munich: C. H. Beck, 1991.

Buchez, Philippe. "De la physiologie." *Le producteur* 3 (1826): 122–23, 264–80, 459–78.

Buffon, Georges Louis Leclerc, compte de. *Natural History, General and Particular, by the Count de Buffon*. Vol. 2. Translated by William Smellie. London: W. Strahan and T. Cadell, 1785. https://books.google.com/books?id=lH0rAAAAYAAJ.

Buisson, Matthieu-François-Régis. *De la division la plus naturelle des phénomènes physiologiques considérés chez l'homme, avec un précis historique sur M. F. X. Bichat*. Paris: Brosson, 1802.

Bülow, Hans von. *Lehrsätze des neuern Krieges oder reine und angewandte Strategie aus dem Geist des neuern Kriegssystems hergeleitet*. Berlin: Heinrich Frölich, 1805.

Burke, Peter. *The Fortunes of the Courtier: The European Reception of Castiglione's Cortegiano*. London: Polity, 1995.

Buytendijk, Frederik J. J. *Allgemeine Theorie der menschlichen Haltung und Bewegung als Verbindung und Gegenüberstellung von physiologischer und psychologischer Betrachtungsweise*. Berlin: Springer, 1972 [1948].

Georges Canguilhem, "Aspects of Vitalism." In *Knowledge of Life*, translated by S. Geroulanos and D. Ginsburg), 59–74. New York: Fordham University Press, 2008.

Canguilhem, Georges. *La connaissance de la vie*. 2. ed. Paris: Vrin, 1965.

Carlet, Gaston. *Essai expérimental sur la locomotion humaine: Étude de la marche*. Paris: Masson, 1872.

Caron, François. *La dynamique de l'innovation: Changement technique et changement social (XVIe–XXe siècle)*. Paris: Gallimard, 2010.

Casper, Stephen T. "The Patient's Pitch: The Neurologist, the Tuning Fork, and Text-book Knowledge." In *The Neurological Patient in History*, edited by L. Stephen Jacyna and Stephen T. Casper, 21–43. Rochester, NY: University of Rochester Press, 2012.

Caussé, Séverin. *Des empreintes sanglantes des pieds et de leur mode de mensuration*. Toulouse: Chauvin, 1853.

Chadarevian, Soraya de. "Graphical Method and Discipline: Self-Recording Instruments in Nineteenth-Century Physiology." *Studies in History and Philosophy of Science* 24 (1993): 267–91.

[Chambaud, Ménuret de]. "Observateur." In *Encyclopédie ou Dictionnaire raisonné des sciences, des arts et des métiers*. Vol. 11. Edited by Denis Diderot and Jean-Baptiste le Rond d'Alembert, 310. Paris: Briasson, 1751.

Chappey, Jean Luc. *La Société des observateurs de l'homme (1799–1804): Des anthropologues au temps de Bonaparte*. Paris: Société des études robbespierristes, 2002.

Charcot, Jean-Martin. *Clinical Lectures on Certain Diseases of the Nervous System*. Translated by Edward P. Hurd. Detroit: G. S. Davis, 1888.

Charcot, Jean-Martin *Leçons du mardi à la Salpêtrière: Policlinique 1887–1888*. 2nd ed. 2 vols. Paris: Bureaux du progrès medical, 1892.

Charcot, Jean-Martin. *Lectures on the Diseases of the Nervous System*. Translated by George Sigerson. London: New Sydenham Society, 1881.

Chaussier, François. *Table synoptique des propriétés caractéristiques et des principaux phénomènes de la force vitale*. 2nd ed. Paris: Théophile Barrois, 1800.

Clausewitz, Carl von. *Vom Kriege: Hinterlassenes Werk*. 3 vols. Edited by Marie von Clausewitz. Berlin: Ferdinand Dümmler, 1832–1834. English translation from Clausewitz, *On War*, translated and edited by Michael Eliot Howard and Peter Paret. Princeton, NJ: Princeton University Press, 1989.

Colin, Gabriel. "Sur l'importance au point de vue médical des signes extérieurs des fonctions de la vie." *Bulletin de l'Académie de médecine*, 2nd ser., 7 (1878): 626. [Discussion of Marey's lecture]

Colin, Gabriel. *Traité de physiologie comparée des animaux*. Paris: Baillière et fils, 1856.

Combes, Marguerite. *Pauvre et aventureuse bourgeoisie: Roulin et ses amis*. Paris: J. Peyronnet, 1928.

Comte, Charles, and Félix Regnault. "Étude comparative entre la méthode de marche et de course dite de fléxion et les allures ordinaires." *Archives de physiologie* (1896): 380–89.

Comte, Charles, and Félix Regnault. "Marche et course en flexion." *Bulletins de la Société d'anthropologie de Paris* 7, no. 7 (1896): 337–41.

Corbin, Alain. Jean-Jacques Courtine, and Georges Vigarello, eds. *Histoire du corps*. 3 vols. Paris: Le Seuil, 2011.

Coulomb, Charles de. "Résultat de plusieurs expériences destinées à déterminer la quantité d'action que les hommes peuvent fournir par leur travail journalier, suivant les différentes manières dont ils emploient leurs forces." In *Théorie des machines simples, en ayant égard au frottement de leurs parties et à la roideur des cordages*, 255–97. Paris: Bachelier, 1821.

Coulomb, Charles de. *Théorie des machines simples*. Paris: De l'imprimerie de Moutard, 1782.

Coulomb, Charles de. *Théorie des machines simples, en ayant égard au frottement de leurs parties et à la roideur des cordages*. Paris: Bachelier, 1821.

Crary, Jonathan. *Techniques of the Observer*. Cambridge, MA: MIT Press, 1990.

Cunningham, Andrew. *The Anatomist Anatomis'd: An Experimental Discipline in Enlightenment Europe*. Farnham: Ashgate, 2010.

Curnieu, Charles-Louis de. *Leçons de science hippique générale; ou, Traité complet de l'art de connaître, de gouverner et d'élever le cheval*. 3 vols. Paris: Librairie militaire J. Dumaine, 1855–1860.

Curtius, Ernst Robert. *Balzac*. Frankfurt: Fischer, 1985 [1923].

Curtius, Ernst Robert. *Die knieenden Figuren der altgriechischen Kunst*. Berlin: Archäologische Gesellschaft, 1869.

Dagognet, François. *Etienne-Jules Marey: A Passion for the Trace*. Translated by Robert Galeta and Jeanine Herman. New York: Zone Books, 1992 [1987].

Daston, Lorraine. *Eine kurze Geschichte der wissenschaftlichen Aufmerksamkeit*. Munich: Siemens Stiftung, 2001.

Daston, Lorraine. "The Empire of Observation, 1600–1800." In *Histories of Scientific Observation*, edited by Lorraine Daston and Elizabeth Lunbeck, 81–113. Chicago: University of Chicago Press, 2011.

Daston, Lorraine, and Peter Galison. "The Image of Objectivity." *Representations* 40 (1992): 81–128.

Daston, Lorraine, and Peter Galison. *Objectivity*. New York: Zone Books, 2007.

Davin, Félix. "Introduction par Félix Davin aux *Études Philosophiques*." In Honoré de Balzac, *Œuvres completes*, vol. 10, 1199–218. Paris: Gallimard, 1979.

Degérando, Joseph-Marie. *Considération sur les diverses méthodes à suivre dans l'observation des peuples sauvages*. N.p., 1800.

Degérando, Joseph-Marie. *Des signes, et de l'art de penser considérés dans leurs rapports mutuels*. 4 vols. Paris: Chez Guson fils, Fuschs, Henrichs, 1799–1800.

Demenÿ, Georges, and Edouard Quénu. "De la locomotion dans l'ataxie locomotrice." *Comptes rendus des séances de l'Académie des sciences* 108 (1889): 963–64.

Demenÿ, Georges, and Edouard Quénu. "Étude de la locomotion humaine dans les cas pathologiques." *Comptes Rendus des Séances de l'Académie des Sciences* 107 (1888): 1550–64.

Desaguliers, Jean Théophile. *Cours de physique expérimentale*. Paris: Rollin, 1751.

Dias, Nélia. *La mesure des sens: Les anthropologues et le corps humain au XIXe siècle*. Paris: Aubier / Flammarion, 2004.

Dictionnaire encyclopédique des sciences médicales. Edited by Amédée Dechambre, Léon Lereboullet, and Louis Hahn. Paris: Masson, 1876.

Diderot Denis. *Lettre sur les aveugles à l'usage de ceux qui voient: Lettre sur les sourds et muets à l'usage de ceux qui entendent et qui parlent.* Paris: Flammarion, 2000.

Digard, Jean-Pierre. *Une histoire du cheval.* 2nd ed. Arles: Actes Sud, 2007.

Dorikens, Maurice, ed. *Joseph Plateau. 1801–1883: Leven tussen Kunst en Wetenschap.* Gent: Provincie Oost-Vlaanderen, 2001.

Douard, John. "E.-J. Marey's Visual Rhetoric and the Graphic Decomposition of the Body." *Studies in the History and Philosophy of Science* 26, no. 2 (1995): 175–204.

Dubois d'Amiens, Frédéric. "Physiologie médicale didactique et critique par M.P.N. Gerdy." *Revue médicale française et étrangère* 4 (1831): 445–67.

Duchenne de Boulogne, [Guillaume-Benjamin]. *Du second temps de la marche, suivie de quelques déuctions pratiques: Mémoire présenté à l'Académie des sciences.* Paris: l'Union médicale, 1855.

Duchenne de Boulogne, [Guillaume-Benjamin]. *Physiology of Motion.* Translated by Emanuel B. Kaplan. Philadelphia: J. B. Lippincott, 1949.

Duchenne de Boulogne, [Guillaume-Benjamin]. *Physiologie des mouvements.* Paris: J. B. Baillière et fils, 1867.

Duchet, Michèle. *Anthropologie et histoire au siècle des Lumières.* Paris: Albin Michel, 1995 [1971].

Dugès, Antoine. *Traité de physiologie comparée de l'homme et des animaux.* 3 vols. Montpellier / Paris: Castel, 1838.

Duhousset, Emile. "Études sur les Kabyles du Djurjura." *Bulletins de la Société d'anthropologie de Paris* 2, no. 3 (1868): 265–71.

Duhousset, Emile. *Le cheval: Allures, extérieur, proportions.* Paris: Morel, 1881.

Duhousset, Emile. *Le cheval: Études sur les allures, l'extérieur et les proportions du cheval; Analyse de tableaux représentant des animaux; Dédié aux artistes.* Paris: Chasles, 1874.

Duhousset, Emile. "Proportions comparatives de l'homme et du cheval." *Gazette des Beaux-Arts* 5, no. 3 (1891): 385–400.

Dumas, Charles-Louis. *Discours sur les progrès futurs de la Science de l'homme.* Montpellier: Tournel, 1804.

Dumas, Charles-Louis. "Observation sur le squelette d'un sauteur, dont les membres abdominaux (extrémités inférieures), étoient composés d'une seule pièce et du pied: suivie de quelques réflexions sur la théorie du saut." *Recueil périodique de la Société de médecine de Paris* 10 (1800): 30–35.

Dumas, Charles-Louis. *Principes de physiologie.* 2nd ed. Vol. 4. Paris: Méquignon, 1806.

Dumas, Charles-Louis. *Principes de physiologie, ou Introduction à la science expérimentale, philosophique et médicale de l'homme vivant.* 4 vols. Paris: Deterville, 1800.

Dupin, Carl. *Geometrie und Mechanik der Künste und Handwerke und der schönen Künste.* 3 vols. Paris: Levrault, 1826.

Eisenberg, Christiane. *"English Sports" und Deutsche Bürger: Eine Gesellschaftsgeschichte 1800–1939.* Paderborn: Ferdinand Schöningh, 1999.

Elliott, Paul. "Vivisection and the Emergence of Experimental Physiology in

Nineteenth-Century France." In *Vivisection in Historical Perspective*, edited by Nicolaas Ruppke, 48–77. London: Routledge, 1987.

Elvert, Jürgen, and Michael Salewski, eds. *Militär, Musik und Krieg—Kolloquium anlässlich des 70. Geburtstages von Michael Salewski*. Historische Mitteilungen der Ranke-Gesellschaft 22. Stuttgart: Franz Steiner, 2009.

Emmanuel, Maurice. *Essai sur l'orchestique grecque: Étude de ses mouvements d'après les monuments figures*. Paris: Hachette, 1895.

Engel, Johann Jakob. *Ideen zu einer Mimik: Erster Theil*. Berlin: Mylius, 1785.

Esquirol, Étienne. "Démonomanie." In *Dictionnaire des sciences médicales*. Vol. 8, 294–318. Paris: Panckoucke, 1814.

Esquirol, Étienne. "Manie." In *Dictionnaire des sciences médicales*. Vol. 30, 437–72. Paris: Panckoucke, 1818.

Exner, Sigmund. *Physiologisches und Pathologisches in den bildenden Künsten*. Vienna: Verein zur Verbreitung naturwissenschaftlicher Kenntnisse, 1889.

Favre, Alphonse. *H.-B. de Saussure et les Alpes*. Lausanne: Bridel, 1870.

Fellows, Otis. "Buffon and Rousseau: Aspects of a Relationship." *Publications of the Modern Language Association of America* 75, no. 3 (1960): 184–96.

Fischer, Otto. "Der Gang des Menschen V. Theil: Die Kinematik des Beinschwingens." *Abhandlungen der mathematisch-physischen Classe der königlich sächsischen Gesellschaft der Wissenschaften* 28, no. 5 (1903): 321–418.

Fischer, Otto. "Der Gang des Menschen VI. Theil: Über den Einflusz der Schwere und der Muskeln auf die Schwingungsbewegung des Beins." *Abhandlungen der mathematisch-physischen Classe der königlich sächsischen Gesellschaft der Wissenschaften* 28, no. 7 (1904): 533–617.

Flesher, Mary Mosher. "Repetitive Order and the Human Walking Apparatus: Prussian Military Science versus the Webers' Locomotion Research." *Annals of Science* 54, no. 5 (1997): 463–87.

Frieling, Kirsten O. "Haltung bewahren: Der Körper im Spiegel frühneuzeitlicher Schriften über Umgangsformen." In *Bewegtes Leben: Körpertechniken in der frühen Neuzeit*, edited by Rebekka von Mallinckrodt, 39–59. Wolfenbüttel: Herzog August Bibliothek, 2008.

Frizot, Michel. *Etienne-Jules Marey: Chronophotographe*. Paris: Nathan, 2001.

Forster-Hahn, Françoise. "Marey, Muybridge and Meissonier: The Study of Movement in Science in Art." In *Eadweard Muybridge: The Stanford Years, 1872–1882*, 85–109. Stanford, CA: Stanford University Museum of Art, 1972.

Foucault, Michel. *The Birth of the Clinic: Archaeology of Medical Perception*. Translated by Alan Sheridan. London: Routledge, 1989 [1963].

Foucault, Michel. *Discipline and Punish: The Birth of the Prison*. Translated by Alan Sheridan. New York: Vintage Books, 1995 [1977].

Garb, Tamar. *Bodies of Modernity: Figure and Flesh in Fin-de-Siècle France*. London: Thames & Hudson, 1998.

Gastine, Louis. *La chronophotographie sur plaque fixe et sur pellicule mobile*. Paris: Gauthier-Villars / Masson, 1897.

Gauchet, Marcel, and Gladys Swain. *La pratique de l'esprit humain: L'institution asilaire et la révolution démocratique*. Paris: Gallimard, 1980.

Gautier, Théophile. "Artistes contemporains: Meissonier." *Gazette des Beaux-Arts*, 1st ser., May 12, 1862, 427–28.

Gerdy, Pierre-Nicolas. *Anatomie des formes extérieures du corps humain, appliquée à la peinture, à la scultpure et à la chirurgie.* Paris: Béchet jeune, 1829.

Gerdy, Pierre-Nicolas. "Discussion sur les fonctions du système nerveux." *Bulletin de l'Académie Royale de Médecine* 3 (1838/39): 415.

Gerdy, Pierre-Nicolas. *Essai de classification naturelle et d'analyse des phénomènes de la vie.* Paris: J.-B. Baillière, 1823 [1821].

Gerdy, Pierre-Nicolas. "Mémoire sur le mécanisme de la marche de l'homme." *Journal de physiologie* 9 (1829): 1–28.

Gerdy, Pierre-Nicolas. *Physiologie philosophique des sensations et de l'intelligence fondée sur des recherches et des observations nouvelles et applications à la morale, à l'éducation, à la politique.* Paris: Labé, 1846.

Gerdy, Pierre-Nicolas. *Physiologie médicale, didactique et critique*, vol. 1, pt. 1, viii–ix. Paris: Crochard, 1832 [1830].

Gérome, Jean-Louis. "Notes & Fragments des Souvenirs inédits du Maître." *Les arts* (1904): 30.

Gerstner, Franz Joseph Ritter von. *Handbuch der Mechanik, aufgesetzt, mit Beiträgen von neuern englischen Konstruktionen vermehrt und herausgegeben von Franz Anton Ritter von Gerstner.* 2nd ed. 3 vols. Prague: Johann Spurny, 1833.

Giedion, Sigfried. *Mechanization Takes Command: A Contribution to Anonymous History.* New York: Oxford University Press, 1948.

Gilles de la Tourette, Georges. *Études cliniques et physiologiques sur la marche: La marche dans les maladies du système nerveux étudiée par la méthode des empreintes; Thèse pour le doctorat en médecine présentée et soutenue le lundi 28 décembre 1885 à 2 heures.* Paris: Imprimerie de la Société de typographie, 1885.

Gilles de la Tourette, Georges. "L'attitude de la marche dans l'hémiplégie hystérique." *Nouvelle iconographie de la Salpêtrière* 1 (1888): 1–12.

Ginzburg, Carlo. "Clues: Roots of an Evidential Paradigm." In *Clues, Myths, and the Historical Method*, 96–125. Baltimore: Johns Hopkins University Press, 1989.

Ginzburg, Carlo. "Family Resemblances and Family Trees: Two Cognitive Metaphors." *Critical Inquiry* 30, no. 3 (Spring 2004): 537–56.

Ginzburg, Carlo "Réflexions sur une hypothèse." In *Mythes, emblèmes, traces: Morphologie et histoire*, 351–64. Lagrasse: Verdier, 2010.

Giraud-Teulon, Félix. "Locomotion." In *Dictionnaire encyclopédique des sciences médicales*, 2nd ser., edited by Amédée Dechambre, Léon Lereboullet, and Louis Hahn, vol. 2, 786–87. Paris: Asselin / Masson et fils, 1869.

Giraud-Teulon, Félix. "Mécanique animale: Étude retrospective sur les progrès realisés dans l'histoire des mécanismes de la locomotion chez l'homme, par la méthode des inscriptions graphiques de M. le professeur Marey." *Bulletin de l'Académie de médecine*, September 4, 1883, 1029.

Giraud-Teulon, Félix. *Principes de mécanique animale; ou, Étude de la locomotion chez l'homme et les animaux vertébrés.* Paris: J.-B. Ballière et fils, 1858.

Goetz, Christopher G., Michel Bonduelle, and Toby Gelfand. *Charcot: Constructing Neurology.* Oxford: Oxford University Press, 1995.

Goetz, Christopher G. *Charcot, the Clinician: The Tuesday Lessons; Excerpts from Nine Case Presentations on General Neurology Delivered at the Salpêtrière Hospital in 1887–88 by Jean-Martin Charcot.* New York: Raven Press, 1987.

Goldstein, Jan. *Console and Classify: The French Psychiatric Profession in the Nineteenth Century.* Cambridge: Cambridge University Press, 1987.

Gooday, Graeme. "Placing or Replacing the Laboratory in the History of Science?" *Isis* 99 (2008): 783–95.

Gordon, Alexander. *Historische und practische Abhandlung über Fortbewegung ohne Thierkraft mittelst Dampfwagen auf gewöhnlichen Landstraßen.* Weimar: Landes-Industrie-Comptoirs, 1833 [1832].

Gotlieb, Marc J. *The Plight of Emulation: Ernest Meissonier and French Salon Painting.* Princeton, NJ: Princeton University Press, 1996.

Gowers, William. *A Manual of Diseases of the Nervous System.* 2 vols. London: Churchill, 1886–1888.

Gray, Thomas. *Observations on a General Iron Railway, or Land Steam-Conveyance: To Supersede the Necessity of Horses in All Public Vehicles; Showing Its Vast Superiority in Every Respect, over All the Present Pitiful Methods of Conveyance by Turnpike Roads, Canals, and Coasting-Traders.* 5th ed. London: Baldwin, Cradock, and Joy, 1825.

Gréard, Octave. *Jean-Louis-Ernest Meissonier: Ses souvenirs—ses entretiens.* Paris: Hachette, 1897.

Greene, Ann Norton. *Horses at Work: Harnessing Power in Industrial America.* Cambridge, MA: Harvard University Press, 2008.

Griep, Wolfgang. "Reiseliteratur im späten 18. Jahrhundert." In *Deutsche Aufklärung bis zur Französischen Revolution: 1680–1789*, edited by Rolf Grimminger, 739–64. Munich: Hanser, 1980.

Griep, Wolfgang, and Hans-Wolf Jäger, eds. *Reise und soziale Realität am Ende des 18. Jahrhunderts.* Heidelberg: Winter, 1983.

Grolle, Joist. "Republikanische Wanderungen: Die Fußreisen des Jonas Ludwig von Heß aus Hamburg durch die 'Freien deutschen Reichsstädte' 1789–1800." *Zeitschrift des Vereins für Hamburgische Geschichte* 83, no. 1 (1997): 299–321.

Gros, Frederic. *A Philosophy of Walking.* London: Verso, 2015 [2009].

Gross, Hans. *Handbuch der Kriminalistik.* 8th ed. Berlin: J. Schweitzer, 1908.

Gross, Michael. "The Lessened Locus of Feeling: A Transformation in French Physiology in the Early Nineteenth Century." *Journal of the History of Biology* 12, no. 2 (1979): 231–71.

Guibert, Jacques-Antoine-Hippolyte de. *Essai général de tactique, précédé d'un discours sur l'état actuel de la politique et de la science militaire en Europe; avec le plan d'un ouvrage intitulé: La France politique et militaire.* 2 vols. London: Chez les libraires associés, 1772.

Guibert, Jacques-Antoine-Hippolyte de. *Versuch über die Tactik. Nebst einer vorläufigen Abhandlung über den gegenwärtigen Zustand der Staats- und Krieg-Wissenschaft in Europa und dem Entwurf eines Werks, betitelt: Das politische und militärische Frankreich.* 2 vols. Dresden: Walthersche Hofbuchhandlung, 1774.

Gunthert, André. "Klinik des Sehens: Albert Londe, Wegbereiter der medizinischen Fotografie." *Fotogeschichte* 21 (2001): 27–40.

Gutsmuths, Johann Christoph Friedrich. *Gymnastik für die Jugend.* 2nd ed. Schnepfenthal: Buchhandlung der Erziehungsanstalt, 1804.

Hadot, Pierre. *Le voile d'Isis: Essai sur l'histoire de l'idée de la nature.* Paris: Gallimard, 2004.

Hadot, Pierre. *The Veil of Isis: An Essay on the History of the Idea of Nature.* Translated by Michael Chase. Cambridge, MA: Harvard University Press, 2006.

Harrington, Anne. *Reenchanted Science: Holism in German Culture from Wilhelm II to Hitler.* Princeton, NJ: Princeton University Press, 1996.

Hecht, Jennifer. *The End of the Soul: Scientific Modernity, Atheism, and Anthropology in France.* New York: Columbia University Press, 2005.

[Heß, Jonas Ludwig von]. *Durchflüge durch Deutschland, die Niederlande und Frankreich.* 3 vols. Hamburg: Bachmann und Gundermann, 1793–1800.

Heydebreck, Adolf von. "Über die Grenzen von Malerei und Plastik." *Verhandlungen der philosophischen Gesellschaft zu Berlin* 7/8 (1878): 1–41.

Hodak, Caroline. "Créer du sensationnel: Spirale des effets et réalisme au sein du theatre équestre vers 1800." *Terrain* 46 (March 2006): 49–66.

Hodak, Caroline. "Du spectacle militaire au théâtre équestre." In *Le cheval et la guerre: Du XVème au XXème siècle,* edited by Daniel Roche, 367–77. Versailles: Association pour l'académie d'art équestre de Versailles, 2002.

Hodak, Caroline. "Du théâtre équestre au cirque: 'une entreprise si éminemment nationale'; Commercialisation des loisirs, diffusion des savoirs et théâtralisation de l'histoire en France et en Angleterre (1760–1860)." Diss., École des hautes études en sciences sociales, 2004.

Hoffmann, Christoph. *Unter Beobachtung: Naturforschung in der Zeit der Sinnesapparate.* Göttingen: Wallstein, 2006.

Hogarth, William. *The Analysis of Beauty.* London: J. Reeves, 1753.

Holmes, Frederic L., and Kathryn M. Olesko. "The Images of Precision: Helmholtz and the Graphical Method in Physiology." In *The Values of Precision,* edited by M. Norton Wise, 198–221. Princeton NJ: Princeton University Press, 1995.

Horath, Daniel. "Die Beherrschung des Krieges in der Ordnung des Wissens: Zur Konstruktion und Systematik der militairischen Wissenschaften im Zeichen der Aufklärung." In *Wissenssicherung, Wissensordnung und Wissensverarbeitung,* edited by Theo Stammen and Wolfgang Weber, 371–86. Berlin: Akademie, 2004.

Horath, Daniel. "Spätbarocke Kriegspraxis und aufgeklärte Kriegswissenschaften: Neue Forschungen und Perspektiven zu Krieg und Militär im 'Zeitalter der Aufklärung.'" *Aufklärung* 12 (2000): 5–47.

Horner, William George. "On the Properties of the Daedaleum, a New Instrument of Optical Illusion." *Philosophical Magazine* 4 (1834): 36–41.

Hugoulin, M. "Reproduction des empreintes des pas, de coups de fusil, etc. sur la neige en matière criminelle." *Annales d'hygiène publique et de médecine légale,* n.s., 3 (1855): 207–12.

Hugoulin, M. "Solidification des empreintes des pas sur les terrains les plus meubles en matière criminelle." *Annales d'hygiène publique et de médecine légale* 44 (1850): 429–32.

Humboldt, Alexander von. *Alexander von Humboldt: Vier Jahrzehnte Wissenschafts-förderung; Briefe an das preußische Kultusministerium 1818–1859.* Edited by Kurt Biermann. Berlin: Akademie 1985.

Humboldt, Alexander von. *Über einen Versuch den Gipfel des Chimborazo zu ersteigen: Mit dem vollständigen Text des Tagebuches "Reise zum Chimborazo."* Edited by Oliver Lubrich and Ottmar Ette. Frankfurt: Eichborn, 2006.

Humbold, Alexander von. "Ueber zwei Versuche den Chimborazo zu besteigen." In *Jahrbuch für 1837*, edited by H. C. Schumacher, 176–206. Stuttgart: Cotta, 1837.

Humphrey, George. *The Story of Man's Mind.* Boston: Small, Maynard, 1923.

Jacyna, Stephen L. "*Medical* Science and *Moral* Science: The Cultural Relations of Physiology in Restoration France." *History of Science* 25, no. 2 (1987): 111–46.

Jahn, Friedrich Ludwig. *Deutsche Turnkunst: Zum zweiten Male und sehr vermehrt herausgegeben.* Berlin: Reimer, 1847.

Jahn, Friedrich Ludwig, and Ernst Wilhelm Eiselen. *Die Deutsche Turnkunst zur Einrichtung der Turnplätze.* Berlin: printed by the author, 1816.

Jähns, Max. *Geschichte der Kriegswissenschaften vornehmlich in Deutschland.* 3 vols. Munich: R. Oldenbourg, 1891.

Jaumes, Alphonse. "Étude des procédés employés pour relever les empreintes sur le sol." *Annales d'hygiène publique et de médecine légale* ser. 3, 3 (1880): 168–77.

Kalof, Linda, and William Bynum, eds. *A Cultural History of the Human Body.* 6 vols. London: Blackwell, 2010.

Kaschuba, Wolfgang. "Die Fußreise: Von der Arbeitswanderung zur bürgerlichen Bildungsbewegung." In *Reisekultur: Von der Pilgerfahrt zum modernen Tourismus*, edited by Hermann Bausinger, Klaus Beyrer, and Gottfried Korff, 165–73. Munich: C. H. Beck, 1991.

Kehlmann, Daniel. *Die Vermessung der Welt.* Reinbek: Rowohlt, 2005.

King, Lester Snow. *The Philosophy of Medicine: The 18th Century.* Cambridge, MA: Harvard University Press, 1978.

Kirchner, Martin. *Grundriss der Militärgesundheitspflege.* Braunschweig: Bruhn, 1896.

Kirchner, Thomas. "Chodowiecki, Lavater und die Physiognomiedebatte in Berlin." In *Daniel Chodowiecki (1726–1801): Kupferstecher, Illustrator, Kaufmann*, edited by Ernst Hinrichs and Klaus Zernack, 101–42. Tübingen: Niemeyer, 1997.

Kleinschmidt, Harald. *Tyrocinium militare: Militärische Körperhaltungen und -bewegungen im Wandel zwischen dem 14. und dem 18. Jahrhundert.* Stuttgart: Autorenverlag, 1989.

Knott, Robert. "Weber, Wilhelm." *Allgemeine Deutsche Biographie* 41 (1896): 358–61.

Kohler, Robert. "Lab History," *Isis* 99 (2008): 761–68.

Kohler, Robert. *Landscapes and Labscapes: Exploring the Lab-Field Border in Biology.* Chicago: University of Chicago Press, 2002.

König, Gudrun M. *Eine Kulturgeschichte des Spaziergangs: Spuren einer bürgerlichen Praktik 1780–1850.* Cologne: Böhlau, 1996.

Krebs, Heinrich Johannes. *Anfangsgründe der eigentlichen Kriegswissenschaft: Aus den besten militärischen Schriften zusammengetragen.* Leipzig: Korte, 1784.

Krüger, Gustav, ed. *Zur Erinnerung an Gerhard Anton Ulrich Vieth, weiland Schulrat und Direktor der Herzogl. Hauptschule zu Dessau. 1786–1836. Aus seinem Nachlass.* Dessau: Paul Baumann, 1885.

Krüger, Johann Gottlob. *Naturlehre. Zweyter Theil, welcher die Physiologie, oder Lehre von dem Leben und der Gesundheit der Menschen in sich fasset.* Halle: Hermann Hemmerde, 1748.

Krünitz, Johann Georg. *Ökonomisch-technologische Encyklopädie.* Berlin: J. Pauli, 1773–1858. http://www.kruenitz1.uni-trier.de.

Kusukawa, Sachiko. "The Uses of Pictures in the Formation of Learned Knowledge: The Cases of Leonhard Fuchs and Andreas Vesalius." In *Transmitting Knowledge: Words, Images, and Instruments in Early Modern Europe,* edited by Sachiko Kusukawa and Ian MacLean, 73–96. Oxford: Oxford University Press, 2006.

Kwint, Marius. "The Circus and Nature in Late Georgian England." In *Histories of Leisure: Leisure, Consumption, and Culture,* edited by Rudy Koshar, 45–60. Oxford: Bloomsbury, 2002.

Kwint, Marius. "The Legitimation of the Circus in Late Georgian England." *Past and Present* 174 (February 2002): 72–115.

Lafontaine, Louis-Firmin [Firmin Weiss]. *Manuel-théorie des courses à pied.* Sens: Modrine, 1888.

Laisné, Napoleon. *Gymnastique pratique, contenant la description des exercices, la construction et le prix des machines, et des chants spéciaux inédits: Ouvrage destiné aux familles, aux établissements d'éducation, aux corps militaires.* Paris: Dumaine, 1850.

Lalanne, Léon. "Méthodes graphiques: Note sur un nouveau mode de représentation de la marche des trains sur une voie de communication." *Comptes rendus des séances de l'Académie des sciences* 99 (1884): 307–13.

Lankester, E. Ray. "The Muybridge Photographs." *Nature* 40 (May 23, 1889): 78–80.

Latocnaye, Jacques-Louis de. *Meine Fußreise durch Schweden und Norwegen: Ein Seitenstück zu der Reise des Verfassers durch die drey brittischen Königreiche; Mit Anmerkungen und Zusätzen eines Deutschen.* 2 vols. Translated by Eduard Henke. Leipzig: Hartknoch, 1802.

Latour, Bruno. "Drawing Things Together." In *Representation in Scientific Practice,* edited by Michael Lynch and Steve Woolgar, 19–68. Cambridge, MA: MIT Press, 1990.

Latour, Bruno. *The Pasteurization of France.* Translated by Alan Sheridan and John Law. Cambridge, MA: Harvard University Press, 1988.

Lauster, Martina. *Sketches of the Nineteenth Century: European Journalism and Its Physiologies, 1830–50.* Palgrave: Macmillan, 2007.

Lauster, Martina. "Walter Benjamin's Myth of the Flâneur." *Modern Language Review* 102 (2007): 139–56.

Lavater, Johann Caspar. *Physiognomische Fragmente zur Beförderung der Menschenkenntnis und Menschenliebe: Vierter Versuch.* 4 vols. Leipzig: Weidmanns Erben und Reich, 1775–1778.

Lavater, Johann Caspar. *Vermischte physiognomische Regeln: Ein Manuscript für Freunde.* Zurich: Orell, Füssli, 1802 [1789].

Lavater, Johann Caspar. *Von der Physiognomik.* Leipzig: Weidemanns Erben und Reich, 1772.

Lawrence, Christopher. "Incommunicable Knowledge: Science, Technology and the Clinical Art in Britain 1850–1914." *Journal of Contemporary History* 20 (1985): 503–20.

Le Bon, Gustave. *L'equitation actuelle et ses principes: Recherches expérimentales.* Paris: Firmin-Didot, 1892.

Lees, A. "Georges Gilles de la Tourette: The Man and His Times." *Revue neurologique* 142, no. 11 (1986): 808–16.

Lefebvre, Thierry, Jacques Malthête, and Laurent Mannoni, eds. *Lettres d'Etienne-Jules Marey à Georges Demenÿ 1880–1894.* Paris: Association française de recherche sur l'histoire du cinéma, 1999.

Legros, Victor. *Étude expérimentale de la marche: Section I; Marche d'un homme isolé.* Paris: Tanera, 1876.

Lenoble Du Teil, Jules. *Étude sur la locomotion quadrupède.* Paris: J. Dumaine, 1873.

Lenoble Du Teil, Jules. *Examen de l'art equestre de M. E. Barroil (Doctrine du Capitaine Raabe): Avec rectification de ses théories, d'après la méthode expérimentale.* Paris: Bureaux de la Revue des Haras, 1890.

Lenoble Du Teil, Jules. *Locomotion quadrupède étudiée sur le cheval.* Paris: Bureau du Journal des Haras, 1877.

Lesch, John. "The Paris Academy of Medicine und Experimental Science, 1820–1848." In *The Investigative Enterprise: Experimental Physiology in Nineteenth-Century Medicine,* edited by William Coleman and Frederic L. Holmes, 100–37. Berkeley: University of California Press, 1988.

Lesch, John. *Science and Medicine in France: The Emergence of Experimental Physiology 1790–1855.* Cambridge, MA: Harvard University Press, 1984.

Leupold, Jacob. "Von den Wagen-Instrumenten." In *Theatri Machinarum Supplementum. Das ist: Zusatz zum Schauplatz der Machinen und Instrumenten,* 22–28. Leipzig: Breitkopf, 1739.

Lévi-Strauss, Claude. *Introduction to the Work of Marcel Mauss.* Translated by Felicity Baker. London: Routledge & Kegan Paul, 1987 [1950].

Lévi-Strauss, Claude. "Jean-Jacques Rousseau, fondateur des sciences de l'homme." In *Anthropologie structurale deux,* 45–56. Paris: Plon, 1973 [1962].

Lévi-Strauss, Claude. "Jean-Jacques Rousseau, Founder of the Sciences of Man." In *Structural Anthropology,* vol. 2, trans. Monique Layton, 33–43. Chicago: University of Chicago Press, 1976.

Le Yaouanc, Moïse. *Nosographie de l'humanité balzacienne.* Paris: Maloine, 1959.

Leys, Ruth. *Trauma: A Genealogy.* Chicago: University of Chicago Press, 1999.

Lichtenberg, Georg Christoph. "Natürliche und affektierte Handlungen des Lebens: Erste Folge." In *Der Fortgang der Tugend und des Lasters. Daniel Chodowieckis Monatskupfer zum Göttinger Taschenkalender mit Erklärungen Georg Christoph Lichtenbergs 1778–1783.* Berlin: Der Morgen, 1977.

Lichtenberg, Georg Christoph. *Schriften und Briefe.* 4 vols. Edited by Wolfgang Promies. Munich: Hanser, 1972.

Lombroso, Cesare. *Criminal Man*. Translated by Mary Gibson and Nicole Hahn Rafter. Durham, NC: Duke University Press, 2006.

Lüh, Jürgen. *Kriegskunst in Europa, 1650–1800*. Cologne: Böhlau, 2004.

MacShane, Clay, and Joel A. Tarr, eds. *The Horse in the City: Living Machines in the Nineteenth Century*. Baltimore: Johns Hopkins University Press, 2007.

Magendie, François. "Note sur les fonctions des corps striés et des tubercules quadri-jumeaux." *Journal de physiologie expérimentale et pathologique* 3, no. 4 (1823): 376–81.

Magendie, François. *Précis élémentaire de physiologie*. Paris: Méquignon-Marvis, 1816–1817.

Magendie, François. *Précis élémentaire de physiologie*, 2nd ed. Paris: Méquignon-Marvis, 1825.

Magendie, François. "Quelques idées générales sur les phénomènes particuliers aux corps vivans." *Bulletin des sciences médicales* 4, no. 24 (1809): 145–70.

Maissiat, Jacques. *Études de physique animale*. Paris: Béthune et Plon, 1843.

Mandressi, Rafael. *Le regard de l'anatomiste: Dissections et invention du corps en Occident*. Paris: Seuil, 2003.

Mannoni, Laurent. *Etienne-Jules Marey: La mémoire de l'œil*. Milan: Mazzotta; Paris: Cinémathèque française, 1999.

Mannoni, Laurent. *The Great Art of Light and Shadow: Archaeology of the Cinema*. Translated and edited by Richard Crangle. Exeter: University of Exeter Press, 2006.

Manouvrier, Léonce. "Étude sur la rétroversion de la tête du tibia et l'attitude humaine à l'époque quaternaire." *Mémoires de la Société d'anthropologie de Paris* 4 (1893): 219–64.

Manouvrier, Léonce. "La platycnémie chez l'homme et chez les singes." *Bulletins de la Société d'anthropologie de Paris* 10 (1887): 128–41.

Marey, Étienne-Jules. *Animal Mechanism: A Treatise on Terrestrial and Aerial Locomotion*. London: King, Bradbury, Agnew, 1874; New York: Appleton, 1879.

Marey, Étienne-Jules. "Des forces utiles dans la locomotion (Conférence faite au congrès international d'hygiène de La Haye, septembre 1884)." *Revue scientifique* 34 (1884): 515–16.

Marey, Étienne-Jules. *Die Chronophotographie*. Translated by Adolf von Heydebreck. Berlin. Mayer & Müller, 1893.

Marey, Étienne-Jules. "Études pratiques sur la marche de l'homme." *La Nature*, January 24, 1885, 119.

Marey, Étienne-Jules. "La chronophotographie et les sports athlétiques." *La Nature*, April 13, 1901, 310–15.

Marey, Étienne-Jules. *La machine animale: Locomotion aérienne et terrestre*. Paris: Librairie Germer Ballière, 1873.

Marey, Étienne-Jules. *La méthode graphique*. Paris: Masson, 1878.

Marey, Étienne-Jules. "La station physiologique." *La Nature*, September 8, 1883, 227.

Marey, Étienne-Jules. *Le movement*. Nimes: J. Chambon, 1994.

Marey, Étienne-Jules. "Moteurs animés." *La Nature*, September 28, 1878, 273.

Mauss, Marcel. "Les techniques du corps." In: *Sociologie et anthropologie*, edited by Claude Lévi-Strauss, 363–72. Paris: Presses universitaires de France, 1950 [1935].

Mauss, Marcel. *Manuel d'ethnographie*. Paris: Payot, 1967 [1947].

Mauss, Marcel. *Techniques, Technology, and Civilization*. Edited by Nathan Schlanger. New York: Berghahn Books, 2006.

Mauss, Marcel. "Techniques of the Body." *Economy and Society* 2, no. 1 (1973): 70–88.

Mayer, Andreas. "Des rêves et des jambes: le problème du corps rêvant (Mourly Vold, Freud, Michaux)." *Romantisme* 178, no. 4 (2017): 75–85.

Mayer, Andreas "Faire marcher les hommes et les images: Les artifices du corps en movement," *Terrain* 46 (March 2006): 33–48.

Mayer, Andreas. "Gradiva's Gait: Tracing the Figure of a Walking Woman." *Critical Inquiry* 38, no. 3 (2012): 554–78.

Mayer, Andreas. *Sites of the Unconscious: Hypnosis and the Emergence of the Psychoanalytic Setting*. Translated by Christopher Barber. Chicago: University of Chicago Press, 2013.

Mégnin, J.-P. *Essai sur les proportions du cheval et son anatomie extérieure comparée à celle de l'homme à l'usage des écuyers militaires ou des artistes*. Paris: Corréard, 1860.

Meissner, Nicolaus N. W. *Geschichte und erklärende Beschreibung der Dampfmaschinen, Dampfschiffe und Eisenbahnen nebst einer Erläuterung der Natur der Wasserdämpfe und der dabei vorkommenden Kunstausdrücke für diejenigen, denen Kenntnisse in Mechanik, Mathematik und Physik fehlen*. Leipzig: G. Fleischer, 1839.

Mekarsky Edler von Menk, Victor. *Das Eisenbahnwesen nach allen Beziehungen kritisch beleuchtet: Für den Gebildeten jeden Standes und ein vollständiges Handbuch für Eisenbahn-Comittéen, Privat-Unternehmer, Mit-Interessenten, Architekten, Ingenieurs und Mechaniker*. Vienna: Franz Tendler, 1837.

Messac, Régis. *Le "detective novel" et l'influence de la pensée scientifique*. Paris: Encrage, 2011 [1929].

Métraux, Alexandre. "Das eidetische Pferd: Wie Bertalan Székely Bildanimationen konstruierte." In *Kunstmaschinen: Spielräume des Sehens zwischen Wissenschaft und Ästhetik*, edited by Andreas Mayer and Alexandre Métraux, 61–100. Frankfurt: Fischer, 2005.

Meyer, Hermann. "Das aufrechte Gehen (Zweiter Beitrag zur Mechanik des menschlichen Knochengerüstes)." *Archiv für Anatomie, Physiologie und wissenschaftliche Medicin* 1853:365–95.

Meyer, Hermann. "Das aufrechte Stehen (Erster Beitrag zur Mechanik des menschlichen Knochengerüstes)." *Archiv für Anatomie, Physiologie und wissenschaftliche Medicin* 1853:9–44.

Meyer, Hermann. "Die Individualitäten des aufrechten Ganges (Vierter Beitrag zur Lehre von der Mechanik des menschlichen Knochengerüstes)." *Archiv für Anatomie, Physiologie und wissenschaftliche Medicin* 1853:548–73.

Meyer, Hermann. *Die richtige Gestalt des menschlichen Körpers in ihrer Erhaltung und Ausbildung für das allgemeine Verständniß dargestellt*. Zurich: Meyer & Zeller, 1874.

Meyer, Hermann. *Die richtige Gestalt der Schuhe: Eine Abhandlung aus der ange-wandten Anatomie für Aerzte und Laien geschrieben.* Zurich: Meyer & Zeller, 1858.

Meyrink, Gustav. "Der Fluch der Krötc – Fluch der Kröte." In *Gesammelte Werke.* Vol. 4, pt. 2. Munich: Langen, 1913.

Mignon, Jacques. *Quelques réflexions sur la mécanique animale, appliquée au cheval.* Paris: Béchet jeune et Labé, 1841.

Miller, Franz von. *Reine Taktik der Infanterie, Cavallerie und Artillerie.* 2 vols. Stuttgart: Buchdruckerei der hohen Karlsschule, 1787–1788.

Momigliano, Arnaldo. "Ancient History and the Antiquarian." *Journal of the Warburg and Courtauld Institutes* 13, no. 3/4 (1950): 285–315.

Montaigne, Michel de. *Essays of Michel de Montaigne.* Translated by Charles Cotton and edited by William Carew Hazlitt. London: Reeves and Turner, 1877.

Montaigne, Michel de. *Les essais.* Paris: Gallimard 2007.

Montègre, Antoine de. "Convulsionnaires." *Dictionnaire des sciences médicales.* Vol. 6, 210–38. Paris: Panckoucke, 1813.

Moravia, Sergio. *Beobachtende Vernunft: Philosophie und Anthropologie in der Aufklärung.* Frankfurt: Fischer, 1989 [1970].

Moravia, Sergio. *La scienza dell'uomo nel Settecento: Con una appendice di testi.* Rome-Bari: Laterza, 1970.

Moravia, Sergio. "Philosophie et médecine en France à la fin du XVIIIe siècle." *Studies in Voltaire and the Eighteenth Century* 39 (1972): 1089–151.

Moritz, Karl Philipp. "Reisen eines Deutschen in England im Jahr 1782." In *Popular-philosophie, Reisen, Ästhetishe Theorie,* vol. 2 of *Werke in zwei Bänden,* edited by Heide Hollmer and Albert Meier, 249–392. Frankfurt: Deutscher Klassiker, 1997.

Morris, Louis Michel. *Essai sur l'extérieur du cheval.* 2nd ed. Paris: Bouchard-Huzard, 1857 [1835].

Müller, Irmgard, and Daniela Watzke. "Weil also die beste Abbildung [. . .] immer ein dürftiges Gleichnis bleibt: Zu den Visualisierungsverfahren im 18. Jahrhundert." In *Anatomie und anatomische Sammlungen im 18. Jahrhundert,* edited by Rüdiger Schultka and Josef N. Neumann, 223–50. Berlin: Hopf, 2007.

Müller, Johannes. *Handbuch der Physiologie des Menschen für Vorlesungen.* 2nd ed. 2 vols. Coblenz: Hölscher, 1840.

Neugebauer, Franz. "Aetiologie der sogenannten Spondylolisthesis." *Archiv für Gynäkologie* 20, no. 3 (1883): 133–84.

Neugebauer, Franz. "Zur Casuistik des sogenannten spondylolisthetischen Beckens." *Archiv für Gynäkologie* 19, no. 3 (1882): 441–74.

Neugebauer, Franz. *Zur Entwickelungsgeschichte des spondylolisthetischen Beckens und seiner Diagnose (mit Berücksichtigung von Körperhaltung und Gangspur).* Dorpat: Niemeyer, 1881.

Newhouse, Ludwig. *Über Chaussee-Dampfwagen, statt Eisenbahnen mit Dampfwagen in Deutschland.* Mannheim: Hoff, 1834.

Nicolai, Friedrich. *Beschreibung einer Reise durch Deutschland und die Schweiz im Jahre 1781. Nebst Bemerkungen über Gelehrsamkeit, Industrie, Religion und Sitten.* 3rd ed. 12 vols. Berlin, 1788.

Noverre, Jean Georges. *Lettres sur la danse et sur les ballets.* Lyon: Aimé Delaroche, 1760.

Ohage, August. "'Raserei für Physiognomik in Niedersachsen': Lavater, Zimmermann, Lichtenberg und die Physiognomik." In *Georg Christoph Lichtenberg 1742–1799: Wagnis der Aufklärung,* 175–84. Vienna: Hanser, 1992.

Olesko, Kathryn. "The Meaning of Precision: The Exact Sensibility in Early Nineteenth-Century Germany." In *The Values of Precision,* edited by M. Norton Wise, 103–34. Princeton, NJ: Princeton University Press, 1995.

Onimus, Ernest. "Des déformations de la plante des pieds, spécialement chez les enfants, dans les affections atrophiques et paralytiques de la jambe: Mémoire lu à l'Association française pour l'avancement des sciences, dans la séance du 19 août 1876." *Gazette hébdomadaire de médecine et de chirurgie* 34 (1876): 531–33.

Onimus, Ernest. "Des déformations du pied et de la jambe." *Revue de chirurgie* 2, no. 6 (1882): 443–62, 652–72.

Onimus, Ernest. "Étude physiologique et clinique des surfaces en contact avec le sol." *Revue de médecine* 1 (1881): 658–59.

Onimus, Ernest, and Charles Viry. *Étude critique des tracés obtenus avec le cardiographe et le sphygmographe.* Paris: Germer Baillière, 1866.

Ory, Pascal. *Le discours gastronomique français.* Paris: Gallimard, 1998.

Osterhammel, Jürgen. *The Transformation of the World: A Global History of the Nineteenth Century.* Princeton, NJ: Princeton University Press, 2014 [2009].

Pagel, Julius Leopold, ed. *Biographisches Lexikon hervorragender Ärzte des neunzehnten Jahrhunderts.* Berlin: Urban & Schwarzenberg 1901.

Pagel, Julius Leopold. "Weber, Eduard Friedrich." *Allgemeine Deutsche Biographie* 41 (1896): 287.

Paul, J. P. "History and Fundamentals of Gait Analysis." *Bio-Medical Materials and Engineering* 8, no. 3/4 (1998): 123–35.

Pellier, Jules. *Le langage équestre.* 2nd ed. Paris: Delagrave, 1900.

Percival, Melissa, and Graeme Tytler, eds. *Physiognomy in Profile: Lavater's Impact on European Culture.* Newark: University of Delaware Press, 2005.

Piechotta, Hans Joachim. "Erkenntnistheoretische Voraussetzungen der Beschreibung: Friedrich Nicolais Reise durch Deutschland und die Schweiz im Jahre 1781." In *Reise und Utopie: Zur Literatur der Spätaufklärung,* edited by Hans Joachim Piechotta, 98–150. Frankfurt: Suhrkamp, 1976.

Piechotta, Hans Joachim, ed. *Reise und Utopie: Zur Literatur der Spätaufklärung.* Frankfurt: Suhrkamp, 1976.

Pinel, Philippe. *Nosographie philosophique, ou la méthode de l'analyse appliquée à la medicine.* Paris: Richard, Caille and Ravier, 1798.

Pire Georges. "Jean-Jacques Rousseau et les relations de voyages." *Revue d'histoire littéraire de la France* 3 (1956): 355–78.

Poivert, Michel. "Variété et vérité du corps humain: L'esthétique de Paul Richer." In *L'art du nu au XIXe siècle: Le photographe et son modèle,* 164–75. Paris: Hazan, 1997.

Preiss, Nathalie. *Les physiologies en France au XIXe siècle: Étude historique, littéraire et stylistique.* Mont-de-Marsan: Éditions InterUniversitaires, 1999.

Punt, Hendrik. *Bernard Siegfried Albinus (1697–1770): On "Human Nature"; Ana-*

tomical and Physiological Ideas in Eighteenth Century Leiden. Amsterdam: B. M. Israël, 1983.

Raabe, Charles. *Examen du Bauchérisme réduit à sa plus simple expression, ou L'art de dresser les chevaux d'attelage, de dame, de promenade, de chasse, de course, d'escadron, de cirque, de tournoi, de carrousel.* Paris: Dumaine, 1857.

Raabe, Charles. *Examen du cours d'équitation de M. Aure, écuyer en chef de l'École de cavalerie, (Saumur 1852).* Marseille: Marius Olive, 1854.

Rabinbach, Anson. *The Human Motor: Energy, Fatigue, and the Origins of Modernity.* Berkeley: University of California Press, 1990.

Regnault, Félix. "Des différentes manières de marcher." *Bulletins de la Société d'anthropologie de Paris* 4 (1893): 586–97.

Regnault, Félix. "Du pas gymnastique." *La Nature,* January 6, 1894, 83–86; January 20, 1894, 122–23.

Regnault, Félix. "Exposition ethnographique de l'Afrique occidentale au Champs-de-Mars à Paris: Sénégal et Soudan français." *La Nature,* August 17, 1895, 183–86.

Regnault, Félix. *Hypnotisme, Religion.* Paris: Reinwald, 1897.

Regnault, Félix. "La chronophotographie dans l'ethnographie." *Bulletins et mémoires de la Société d'anthropologie de Paris,* 5th ser., 1 (1900): 421–22.

Regnault, Félix. "La marche et le pas gymnastique militaires." *La Nature,* July 29, 1893, 129.

Regnault, Félix. "Le grimper." *Revue encyclopédique,* October 23, 1897, 904–5.

Regnault, Felix. "Le langage par le geste." *La Nature,* October 15, 1898, 315–17.

Regnault, Félix. "Méthode de la course en flexion (Dromothérapie)." *Gazette médicale de Paris,* 12th ser., 3, no 43 (October 24, 1903): 349–50.

Regnault, Félix, and Albert-Charlemagne-Oscar de Raoul. *Comment on marche: Des divers modes de progression; De la superiorité du mode en flexion.* Paris: Lavauzelle, 1898.

Reichard, Heinrich August Ottokar. *Der Passagier auf der Reise in Deutschland und einigen angränzenden Ländern, vorzüglich in Hinsicht auf seine Belehrung, Bequemlichkeit und Sicherheit: Ein Reisehandbuch für Jedermann.* Weimar: Gädicke, 1801. https://books.google.com/books?id=xd5QAAAAcAAJ.

Reichard, Heinrich August Ottokar. *Handbuch für Reisende aus allen Ständen.* Leipzig: Weygandtschen Buchhandlung, 1784.

Reinach, Salomon. "Chronique d'Orient." *Revue archéologique* 9 (1887): 107.

Reinach, Salomon. "La représentation du galop dans l'art ancien et modern." *Revue archéologique* 36 (1900): 216–51, 441–50; 37 (1900): 244–59; 38 (1901): 27–45; 39 (1901): 9–11. [2nd ed., Paris, 1925]

Rey, Alain, ed. *Dictionnaire historique de la langue française.* Paris: Le Robert, 1994.

Rey, Roselyne. *Naissance et développement du vitalisme en France de la deuxième moitié du 18e siècle à la fin du Premier Empire.* Oxford: Voltaire Foundation, 2003 [1987].

Richer, Paul. *Canon des proportions du corps humain.* Paris: Delagrave, 1893.

Richer, Paul. *Études sur l'attaque hystéro-épleptique faites à l'aide de la méthode graphique.* Paris: Delahaye, 1879.

Richer, Paul. *Introduction à la figure humaine.* Paris: Gaultier / Magnier, 1902.

Richer, Paul. *Nouvelle anatomie artistique III: Cours supérieur (suite); Physiologie; Attitude et mouvements.* Paris: Plon, 1921.

Richer, Paul. *Physiologie artistique de l'homme en movement.* Paris: Doin, 1895.

Riegl, Alois. "Objektive Ästhetik." *Literatur-Blatt der Neuen Freien Presse,* July 13, 1902.

[Riesbeck, Johann Kaspar]. *Briefe eines reisenden Franzosen über Deutschland an seinen Bruder zu Paris.* 2nd ed. Vol. 1. Zurich, 1784.

Rigoli, Juan. *Lire le délire: Aliénisme, rhétorique et littérature en France au XIXe siècle.* Paris: Fayard, 2001.

Rittershausen. "Die Marschgeschwulst oder das sogenannte Ödem des Mittelfusses." *Militär-Wochenblatt* 75 (1894).

Rivers, Christopher. "'L'homme hieroglyphié': Balzac, Physiognomy, and the Legible Body." In *Faces of Physiognomy,* edited by Ellis Shookman, 144–60. Columbia, SC: Camden House, 1993.

Roche, Daniel. "Equestrian Culture in France from the Sixteenth to the Nineteenth Century." *Past and Present* 199 (May 2008): 113–45.

Roche, Daniel *La culture équestre de l'Occident, XVIe–XIXe siècle: L'ombre du cheval.* 3 vols. Paris: Fayard, 2008–2015.

Roger, Jacques. *Les sciences de la vie dans la pensée française au XVIIe siècle.* Paris: Albin Michel, 1993 [1963].

Rohmer, Joseph. *Les variations de forme normales et pathologiques de la plante du pied étudiées par la méthode graphique.* Nancy: Collin, 1879.

Romberg, Moritz Heinrich. *Lehrbuch der Nervenkrankheiten des Menschen.* 2 vols. Berlin: Duncker, 1840.

Romberg, Moritz Heinrich. *A Manual of the Nervous Diseases of Man.* Translated by Edward H. Sieveking. London: Sydenham Society, 1853.

Rony, Fatimah Tobing. *The Third Eye: Race, Cinema, and Ethnographic Spectacle.* Durham, NC: Duke University Press, 1996.

Roodenburg, Herman. *The Eloquence of the Body: Perspectives on Gesture in the Dutch Republic.* Zwolle: Waanders, 2004.

Rösch, Jacob Friedrich von. *Mathematische Säze aus der Tactik.* Stuttgart: Cotta, 1778.

Rosenfeld, Sophia. *A Revolution in Language: The Problem of Signs in Late Eighteenth-Century France.* Stanford, CA: Stanford University Press, 2001.

Roulin, François-Désiré. *Propositions sur les mouvements et les attitudes de l'homme.* Paris: Didot jeune, 1820.

Roulin, François-Désiré. "Recherches théoriques et expérimentales sur le mécanisme des attitudes et des mouvemens de l'homme." *Journal de physiologie expérimentale* 1, no. 3 (1821): 209–36.

Rousseau, Jean-Jacques. *The Confessions and Correspondence, Including the Letters to Malesherbes.* Translated by Christopher Kelly and edited by Christopher Kelly, Roger D. Masters, and Peter G. Stillman. Hanover: University Press of New England, 1995.

Rousseau, Jean-Jacques. *Discourse on the Origin of Inequality.* Translated by Donald A. Cress. Indianapolis, IN: Hackett, 1992.

Rousseau, Jean-Jacques. *Émile, or Education*. Translated by Barbara Foxley. London: J. M. Dent and Sons; New York: E. P. Dutton, 1921.

Rousseau, Jean-Jacques. *Julie, or The New Heloise: Letters of Two Lovers Who Live in a Small Town at the Foot of the Alps*. Translated by Philip Stewart and Jean Vaché. Dartmouth, NH: Dartmouth College Press, 1997.

Roux, Johann Wilhelm. *Anweisung zum Hiebfechten mit geraden und krummen Klingen*. Jena: Friedrich Mauke, 1840.

Rullier, Pierre Joseph. "Geste." In *Dictionnaire des sciences médicales*. Vol. 18, 329–79. Paris: Panckoucke, 1817.

Rullier, Pierre Joseph. "Locomotion." Vol. 28, 548–78. Paris: Panckoucke, 1818.

Rullier, Pierre Joseph. "Marche." In *Dictionnaire des sciences médicales*. Vol. 31, 6–23. Paris: Panckoucke, 1819.

Rullier, Pierre Joseph. "Mouvement." In *Dictionnaire des sciences médicales*. Vol. 34, 438–59. Paris: Panckoucke, 1819.

Rullier, Pierre Joseph. "Motilité." In *Dictionnaire des sciences médicales*. Vol. 34, 401–3. Paris: Panckoucke, 1819.

[Saldern, Friedrich Christoph von]. *Taktische Grundsätze und Anweisung zu militairischen Evolutionen: Von der Hand eines berühmten Generals*. Dresden: Walthersche Hofbuchhandlung, 1786 [1781].

Saldern, Friedrich Christoph von. *Taktische Grundsätze und Anweisung zu militairischen Evolutionen*. Edited by Heinrich Johannes Krebs. Copenhagen: Johann Heinrich Schubothe, 1796.

Salquin, Samuel Auguste. *Die militärische Fussbekleidung*. Bern: Jent & Reinert, 1881.

Salzmann, Christian Gotthilf. *Ameisenbüchlein, oder Anweisung zu einer vernünftigen Erziehung der Erzieher*. Reutlingen: Mäcken, 1808.

Salzmann, Christian Gotthilf. "Wie gut es sey, seine Kinder das Gehen selbst lernen zu lassen." In *Nachrichten aus Schnepfenthal für Eltern und Erzieher*, vol. 1, 168–73. Leipzig: Siegfried Lebrecht Crusius, 1786.

Sarasin, Philipp, and Jakob Tanner, eds., *Physiologie und industrielle Gesellschaft: Studien zur Verwissenschaftlichung des Körpers im 19. und 20. Jahrhundert*. Frankfurt: Suhrkamp, 1998.

Sarthe, Louis-Jacques Moreau de la, ed. *L'art de connaître les hommes par la physionomie par Gaspard Lavater*. 10 vols. Paris, 1820 [1806–1809].

Saxon, Arthur Hartley. *Enter Foot and Horse: A History of Hippodrama in England and France*. New Haven, CT: Yale University Press, 1968.

Schaffer, Simon. "Self Evidence." *Critical Inquiry* 18, no. 2 (1992): 327–62.

Scheel, Hans Ludwig. "Balzac als Physiognomiker." *Archiv für das Studium der neueren Sprachen und Literaturen* 198 (1962): 227–44.

Schelling, Wilhelm Joseph von. "Von der Weltseele" (1798). In *Friedrich Wilhelm Joseph von Schellings sämmtliche Werke: Erste Abtheilung*. Vol. 2. Stuttgart: Cottascher, 1857.

Schivelbusch, Wolfgang. *The Railway Journey: The Industrialization of Time and Space in the Nineteenth Century*. Berkeley: University of California Press, 1986 [1877].

Schlanger, Nathan. Introduction to *Techniques, Technology, and Civilization*, by Marcel Mauss, 1–29. New York: Berghahn Books, 2006.

Schmitt, Jean-Claude. *La raison des gestes dans l'Occident médiéval*. Paris: Gallimard, 1990.

Schmitt, Jean-Claude. "The Rationale of Gestures in the West: Third to Thirteenth Centuries." In *A Cultural History of Gesture from Antiquity to the Present Day*, edited by Jan Bremmer and Herman Roodenburg, 59–70. Ithaca, NY: Cornell University Press, 1991.

Schmitz, J. W. "Wahrscheinliche Resultate der Dampfwagen auf gewöhnlichen Wegen." *Der Sammler: Ein Unterhaltungsblatt* 85 (July 17, 1834): 340–41.

Schroedter, Stephanie, Marie-Thérèse Mourey, and Gilles Bennett, eds. *Barocktanz im Zeichen französisch-deutschen Kulturtransfers: Quellen zur Tanzkultur um 1700*. Hildesheim: Olms, 2008.

Seume, Johann Gottfried. "Mein Sommer 1805." In *Werke in zwei Bänden*, vol. 1, edited by Jörg Drews. Frankfurt: Deutscher Klassiker, 1993.

Seume, Johann Gottfried. *Spaziergang nach Syrakus im Jahre 1802*. Edited by Albert Meier. 3rd ed. Munich: Deutscher Taschenbuch, 1994.

Shookman, Ellis, ed. *The Faces of Physiognomy: Interdisciplinary Approaches to Johann Caspar Lavater*. Columbia, SC: Camden House, 1993.

Sieburth, Richard. "Une idéologie du lisible: Le phénomène des physiologies." *Romantisme* 47 (1985): 39–60.

Sikora, Michael. "'Ueber die Veredlung des Soldaten': Positionsbestimmungen zwischen Militär und Aufklärung." In *Die Bestimmung des Menschen*, edited by Norbert Hinske, 25–50. Hamburg: Meiner, 1999.

Solnit, Rebecca. *Wanderlust: A History of Walking*. New York: Penguin Books, 2001.

Spazier, Otto. Review of Johann Gottfried Seume, *Spaziergang nach Syracus*. *Zeitung für die elegante Welt*, June 4, 1803.

Stampfer, Simon. *Die stroboskopischen Scheiben oder optischen Zauberscheiben*. Vienna: Trentsensky & Vieweg, 1833.

Starobinski, Jean. *Action and Reaction: The Life and Adventures of a Couple*. New York: Zone, 2003.

Starobinski, Jean. *Histoire du traitement de la mélancolie des origines à 1900*. Basel: Geigy, 1960.

Starobinski. Jean. "Rousseau and Buffon." In *Jean-Jacques Rousseau: Transparency and Obstruction*, translated by Arthur Goldhammer, 323–32. Chicago: University of Chicago Press, 1988.

Staum, Martin. "Cabanis and the Science of Man." *Journal of the History of the Behavioral Sciences* 10, no.1 (1974): 135–43.

Staum, Martin. "Physiognomy and Phrenology at the Paris Athénée." *Journal of the History of Ideas* 56, no. 3 (1995): 443–62.

Steinhausen, Wilhelm. "Mechanik des menschlichen Körpers (Ruhelagen, Gehen, Laufen, Springen)." In *Bewegung und Gleichgewicht, Physiologie der körperlichen Arbeit I: Handbuch der normalen und pathologischen Physiologie*, ed. A. Bethe et al., vol. 15, pt. 1. Berlin: Springer, 1930.

Stewart, William E. *Die Reisebeschreibung und ihre Theorie im Deutschland des 18. Jahrhunderts.* Bonn: Bouvier, 1978.

Stierle, Karl-Heinz. *Der Mythos von Paris: Zeichen und Bewußtsein der Stadt.* Munich: Hanser, 1993.

Stierle, Karl-Heinz. "Epische Naivität und bürgerliche Welt: Zur narrativen Struktur im Erzählwerk Balzacs." In *Honoré de Balzac,* edited by Hans-Ulrich Gumbrecht, Karl-Heinz Stierle, and Rainer Warning, 175–218. Munich: Fink, 1980.

Stübig, Heinz. "Berenhorst, Bülow und Scharnhorst als Kritiker des preußischen Heeres der nachfriderizianischen Epoche." In *Die preußische Armee: Zwischen Ancien Régime und Reichsgründung,* edited by Peter Baumgart, Bernhard R. Kroener, and Heinz Stübig, 107–20. Paderborn: Ferdinand Schönigh, 2008.

Studeny, Christophe. *L'invention de la vitesse: France, XVIIIe–XXe siècle.* Paris: Gallimard, 1995.

Svevo, Italo. *Zenos Gewissen.* Translated by Barbara Kleiner. Frankfurt: Zweitausendeins, 2000.

Taquet, Philippe. *Georges Cuvier: Naissance d'un genie.* Paris: Odile Jacob, 2006.

Taquet, Philippe. "Les premiers pas d'un naturaliste sur les sentiers du Wurtemberg: Récit inédit d'un jeune étudiant nommé Georges Cuvier." *Geodiversilas* 20 (2): 285–318.

Thalwitzer, Franz. *Der Parademarsch: Eine ärztliche Betrachtung.* Dresden: Paul Alicke, 1904.

Thelwall, John, *The Peripatetic.* Edited by Judith Thompson. Detroit: Wayne State University Press, 2001 [1793].

Thiébault-Sisson, François. "Meissonnier [sic]: Ses procédés de travail." *Le Temps,* November 6, 1895.

Thom, Walter. *Pedestrianism; or, An Account of the Performances of Celebrated Pedestrians during the Last and Present Century: With a Full Narrative of Captain Barclay's Public and Private Matches; and an Essay on Training.* Aberdeen: A. Brown and F. Frost, 1813.

Tidy, Charles. *Legal Medicine.* 2 vols. London: Smith, Elder, 1882.

Tolley, Bruce. "Balzac et la doctrine saint-simonienne." *L'année Balzacienne* 1973:159–67.

Tolley, Bruce. "Balzac et les saint-simoniens." *L'année balzacienne* 1966:49–66.

Trocchio, Federico di. "Barthez et l'Encyclopédie." *Revue d'histoire des Sciences* 34, no. 2 (1981): 123–36.

Treitschke, Heinrich von. *Deutsche Geschichte im Neunzehnten Jahrhundert.* New ed. 5 vols. Leipzig: Hendel, 1927 [1879].

Turcot, Laurent. *Le promeneur à Paris au XVIIIe siècle.* Paris: Gallimard, 2007.

Vatin, François. *Le travail et ses valeurs.* Paris: Albin Michel, 2004.

Véron, Pierre. *Les coulisses artistiques.* Paris: Dentu, 1876.

Vierordt, Hermann. *Das Gehen des Menschen in gesunden und kranken Zuständen nach selbstregistrierenden Methoden dargestellt.* Tübingen: H. Laupp'sche Buchhandlung, 1881.

Vieth, Gerhard Ulrich Anton. *Versuch einer Encyclopädie der Leibesübungen.* Edited by Friedrich Fetz. 3 vols. Frankfurt: Limpert, 1970 [1795].

Vigarello, Georges. *Le corps redressé: Histoire d'un pouvoir pédagogique.* Paris: Armand Colin, 2001 [1978].

Vincent, Antoine-François, and Georges-Claude Goiffon. *Mémoire artificielle des principes relatifs à la fidelle représentation des animaux, tant en peinture qu'en sculpture: Première partie concernant le cheval.* Alfort: Vincent and Goiffon, 1779.

Virgil. *Virgil in Two Volumes.* Translated by H. Rushton Fairclough. Cambridge, MA: Harvard University Press, 1960.

Viry, Charles. "De la chaussure du soldat d'infanterie." *Archives de médecine et de pharmacie militaires* 9 (1887): 1–18.

Volkmann, Richard "Die Krankheiten der Bewegungsorgane." In *Handbuch der allgemeinen und speciellen Chirurgie,* edited by Franz von Pitha and Theodor Billroth, vol. 2, 722–27. Erlangen: Enke, 1872.

Wachter, Louis-Rupert. *Aperçus équestres au point de vue de la méthode Baucher.* Paris: Librairie militaire J. Dumaine, 1862.

Wallace, Anne D. *Walking, Literature, and English Culture: The Origins and Uses of Peripatetic in the Nineteenth Century.* Oxford: Clarendon, 1993.

Warneken, Bernd Jürgen. "Biegsame Hofkunst und aufrechter Gang: Körpersprache und bürgerliche Emanzipation um 1800." In *Der aufrechte Gang: Zur Symbolik einer Körperhaltung,* 11–23. Tübingen: Ludwig Uhland Institut, 1990.

Warneken Bernd Jürgen. "Bürgerliche Gehkultur in der Epoche der Französischen Revolution." *Zeitschrift für Volkskunde* 85 (1989): 177–87.

Weber, Eduard. "Einige Bemerkungen über die Mechanik der Gelenke, insbesondere über die Kraft, durch welche der Schenkelkopf in der Pfanne erhalten wird; ein Vortrag gehalten vor der Versammlung der deutschen Naturforscher zu Bonn am 23. September." *Archiv für Anatomie, Physiologie und wissenschaftliche Medicin* 1836:54–59.

Weber, Heinrich. *Wilhelm Weber: Eine Lebensskizze.* Breslau: E. Trewendt, 1893.

[Weber, Wilhelm, and Eduard Weber]. "Mécanique de la locomotion chez l'homme." In *Encyclopédie anatomique II: Ostéologie, syndesmologie, et mécanique des organes locomoteurs,* translated by A.-J.-L. Jourdan, 237–522. Paris: J. B. Ballière, 1843.

Weber, Wilhelm, and Eduard Weber. *Mechanik der menschlichen Gehwerkzeuge. Eine anatomisch-physiologische Untersuchung.* Vol. 6 of *Wilhelm Weber's Werke,* edited by Friedrich Merkel and Otto Fischer. Berlin: Julius Springer, 1894 [1836].

Weber, Wilhelm, and Eduard Weber. "Ueber die Mechanik der menschlichen Gehwerkzeuge, nebst der Beschreibung eines Versuchs über das Herausfallen des Schenkelkopfs aus der Pfanne im luftverdünnten Raum." *Annalen der Physik und Chemie* 40, no.1 (1837): 1–13.

Weiner, Dora, and Michael Sauter. "The City of Paris and the Rise of Clinical Medicine." *Osiris* 18 (2003): 23–42.

Weisbach. "Die sogenannte 'Fussgeschwulst'—Syndesmitis metatarsea— des Infanteristen in Folge von anstrengenden Märschen." *Deutsche militär-ärztliche Zeitschrift* 6 (1877): 551–57.

Wied, Maximilian zu. *Beiträge zur Naturgeschichte von Brasilien.* 4 vols. Weimar: Landes-Industrie-Comptoirs, 1825–1833.

Williams, Elizabeth. *A Cultural History of Medical Vitalism in Enlightenment Montpellier*. Burlington, VT: Ashgate, 2003.

Williams, Elizabeth. *The Physical and the Moral: Anthropology, Physiology, and Philosophical Medicine in France, 1750–1850*. Cambridge: Cambridge University Press, 1994.

Zelle Carsten, ed. *"Vernünftige Ärzte": Hallesche Psychomediziner und die Anfänge der Anthropologie in der deutschsprachigen Frühaufklärung*. Tübingen: Niemeyer, 2001.

Zuntz, Nathan, and Wilhelm August Ernst Schumburg. *Studien zu einer Physiologie des Marsches*. Berlin: August Hirschwald, 1901.